A GUIDE TO HI-FI
AND AUDIO RECORDING

Warwick Thorn

Updated 2024

ISBN: 978-1-7386038-3-1 (Amazon Paperback Edition)
ISBN: 978-1-7386038-1-7 (Draft2Digital Edition)
ISBN: 978-1-7386038-2-4 (Kindle Edition)

TABLE OF CONTENTS

CHAPTER 1
INTRODUCTION

"Quality is never an accident. It is always the result of intelligent effort",
By John Ruskin.

ARE YOU ON A LEARNING CURVE when setting up your audio recording and/or lounge listening spaces? When trying work out the why, the what, the how of acoustics, audio devices and setup options? I answer those questions by integrating all the parts into one holistic system, translated into acoustic treatment, devices, software, setup options, and by bridging between hi-fi listening and AVR, with audio recording usually done in another room. Once you see the system, you will be able to think your way through how you want to do things, unlocking the barriers to discovering your own remarkable production and listening experiences.

I include the **why's**, and the **how to** DIY instructions in cases such as room acoustic treatment, building hi-fi speakers, and more, in many cases providing links to more detailed expert explanations on YouTube. I follow a mid-fi reasonable budget approach, that aims for **what** is needed to create and appreciate clarity and emotional connectivity to audio. Where I recommend gear it is always of quality with good reviews behind it, for a reasonable budget – when estimating cost, I use USD. And, yet I try to edge into high-end. I place a preference somewhere between analog and digital, looking for the best of both worlds; for music and movie listening I look for hybrid analog and digital, and integrated TV and Hi-fi solutions.

Some information pertains to home listening scenarios and other to audio recording. Therefore, to facilitate your possible reading preferences, I use two icons to highlight these two applications:

"Hi-Fi", includes any room, or space in a room, used for listening to music and/or watching TV, and the technology and practices to do so.

"Audio Recording", includes any home studio setup or room, or space in a room, used for recording podcasts and or audio, and/or live on-line or live venue sessions, and the technology and practices to do so.

These are modern times. When our Smart TV broke down, I left it for a while because our family had mostly grown out of shared viewing sessions, sad to say, in a way. We had begun to individually choose our viewing preferences, via TV, smartphones, and computers. Now, my wife mostly likes to read in the same room; while I might listen to music on headphones, perhaps while working, or I might be watching my own movie, while she is watching American Idol on her smartphone and headphones. Or, we enjoy background or focused music, or a movie, together. We have one son still at home, and he tends to bound into the room to share something from his phone on a bigger screen, sound effects, please! My wife doesn't want a TV as the centerpiece of the room, but we sometimes watch a movie together, and these days it's, more often than not, streamed. I swivel our couch around and we watch it on our large enough 4K TV that otherwise doubles up as a mammoth external computer monitor connected to an AVR and Dolby Surround Sound system, which is also interconnected to our Hi-Fi stereo setup. There are many alternative setup options and gear for listening and recording setups. In both cases I try and provide very good and manageable options.

And, I go downstairs to my recording room setup, to conduct online classes, and other recording sessions.

I have a preference for separate devices, rather than integrated. This is because with separate devices you end up with being able to choose the best option for every step rather than being cornered into comparatively limited typology or software.

I also have a preference for modular acoustic treatment. For an audio recording setup, I once built a recording studio in our house, only for us to move and leave it all behind! Now, I have a modular setup of decor friendly acoustic treatment that can be moved to a different room or house as needed, some placed unobtrusively in our lounge. These solutions are explained in this book.

While I focus on audio in this book, I do sometimes discuss the video side of things and how that fits in, because it can be inextricably linked.

We need to understand acoustic treatment, the technology of different devices and how they work as a system, and sometimes the electrical side of it all, and software. **What** is it doing and **why**? And, **how** it works together as a whole system.

While I do not claim to be an expert in all the fields I cover, I do have enough hands on experience, research and teaching skills to develop an integrated way of understanding. There is a significant amount of technical detail, for you to really get to grips with the subject, and make your own decisions with knowledge.

CHAPTER 2
PRODUCED SOUND

"Sound is the vocabulary of nature", by Pierre Schaeffer.

FIRST LET US LOOK AT WHAT MAKES UP SOUND AS IT IS PRODUCED. By understanding this, all the choices we make about audio equipment, where we record, and acoustic treatment, and the best way to listen to music, start to make sense.

Through sound we connect to nature, to the harmonics of the wind in the trees, to the birds that sing. We use it to communicate, not just that, but for our song, which connects and uplift us.

PRODUCED SOUND AND HARMONICS

SOUND IS EXPRESSED IN TWO LAYERS - fundamental frequencies and harmonics. Fundamental frequencies are the first sounds produced, harmonics the extended timbrel or tonal color of that sound, which spreads out beyond the fundamental initial sound. The fundamental sounds are important to record and hear clearly, the harmonics more so affecting tone, which unifies the content emotionally.

Harmonics generate at multiples of the fundamental sound. If you produce a sound at 1000Hz, there will be harmonics of that sound initiated at 2000Hz, 3000Hz, 4000Hz, and so on. Timbrel or tonal color comes from the harmonics reverberating in the enclosed instrument, such as within the shape of a guitar, including the vibrational effect of the wooden material. Likewise with the human throat and nasal cavities, and for that matter, the dimensions of rooms and materials of the reflective surfaces of rooms. Oh, and if you're recording, the microphone itself. Similarly for playback, the playback equipment itself.

Below is a table of musical instruments, including male and female human voices, showing their fundamental sound range, and harmonic upward reach on a frequency scale. The dotted line shows where a subwoofer crossover is usually set, so that before that line the sound generation can be transferred to come from a subwoofer, or the built-in subwoofer of a full range speaker. Beyond that dotted line the sound comes from main drivers and tweeters. It is true, very good main drivers in (usually) large speaker cabinets reach below 80 Hz, but they usually don't do it as clearly as a dedicated subwoofer does. Good headphones can provide bass below 80 Hz, though, to get that right, meaning to avoid one note bass boominess, they need to be high quality headphones, even modified quality headphones (as will be discussed later).

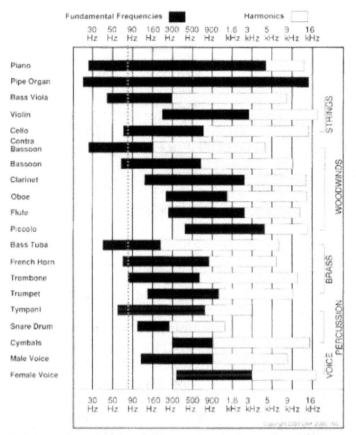

Ref *(researchgate.net/figure/Frequency-ranges-of-several-musical-instruments-30_fig3_228446442)*

Here is a useful video, from Stehlik's Music, on basic frequencies and harmonics, see www.youtube.com/watch?v=WihdMYEmol0.

And, another by Bradly Hamilton, on the uniqueness of tone (timbre), see www.youtube.com/watch?v=heiFFarVU6o.

THE ILLUSTRATION BELOW depicts how fundamental vocal generation covers a frequency range of 200-6000 Hz, mostly focused in 200-4000 Hz.

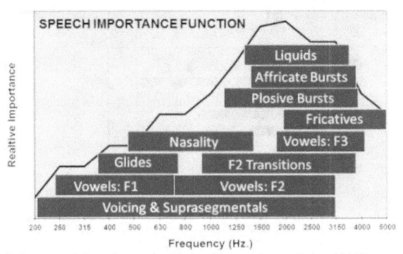

Ref (www.audiologyonline.com/articles/making-speech-more-distinct-12469)

Around the sound range of an idling car engine, Liquids, Affricative and Plosive Bursts, and Fricatives are those precise sounds which make all the difference to understanding what a person is saying, or singing. To be specific, these are: Liquids, /l/ and /r/; Affricative Bursts, 'ch' 'j' 'g'; Plosive bursts, /p/, /t/, and /k/, Fricatives /th/, /h/, /s/, /f/, and /sh/. Voicing & Suprasegmentals are the stress and tonal cues that occur across the widest range, providing an essential second level of emotional communication: implications, hints, sarcasm, eagerness, anger, anguish, delight, lightheartedness, and so on. These are all fundamental sounds of speech – naturally, harmonics reach above that to unify emotional content, more so with singing.

Clearly, these frequencies are critical to understand spoken and sung audio. When an audio engineer mixes other music tracks, those other tracks may need to be scooped out (lowered a bit at these important vocal frequencies, and/or pushed back behind the vocals (achieved with a little reverb/echo). Then for listening, room reverberation, caused by room reflections, and playback technologies and placement can muddy clarity.

FOR MOVIES OR TV, audio includes many common sounds (aka 'effects'), which can, likewise, be understood within certain

8

frequency bands. If we compare these to the vocal frequencies, just discussed, we can understand why it is hard to understand what people are saying when any of these sounds are made:

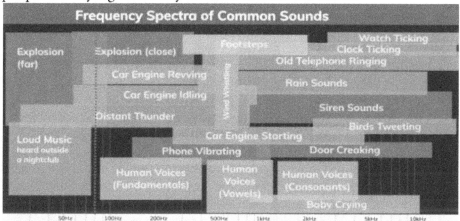

Reference: Source unknown.

These sounds are mostly relevant in terms of TV programs or movie. It may be interesting to note that with Dolby Atmos becoming mainstream, some of those higher frequency effects are moved across the space above your head, thus the spatial placement of the effects is elsewhere from direct vocals (direct vocals will be from the left, right, and center speakers). Atmos above-the-head effects are from those sounds in the higher frequencies, because, for example, it's natural to hear birds tweeting overhead. Whereas, low bass effects will come from your subwoofer, which tends to disperse sound around a room - leaving vocals to be more direct; since our ears naturally distinguish direct from indirect sound.

VIEWING AUDIO IN AUDIO EDITORS

WHEN AUDIO EDITING, there are are two main ways to view audio, front-on and longways. Visualize, for a moment, audio coming through a grey audio cable (image below). We are going to consider the difference of sound front on or long ways, because these are the two ways sound is visualized.

Slice through the end of the grey audio cable, and you see the front-on view, with the spectrum analyzer providing a slice view in real-time of the sound as it is occurring across the frequency range. Since the audio is dynamic, flowing out of the cable, the graph keeps changing dynamically. This spectrum analyzer view is what you see in equalization and compression plug-ins. The left side of the spectrum visualizer is low bass sound, the right side, high notes; high frequency sound. If you lower or raise any frequency area, certain parts of the sound subtly change. With this view in mind, it is common to apply equalization and compression. Equalization is applied to correct over-boominess or shrillness, or it may be used to dip down a certain frequency area so other tracks come through clearly – you can think of each track as a separate cable. Compression plugins are used to reduce the difference between the high and low volumes, so audio is easier to hear in a car where there is background noise.

Now, for the longways, more common, view. If you cut out a window from the top of the cable, you will see the sound from the longways view, for a period of time in one go, usually 10-20 second chunks. This spectrogram view depicts a typical waveform that is used for editing. This view is more useful for editing audio, the blue waveform part usually being the main view of an audio track in a Digital Audio Workstation, and as seen in an audio editor like Izotope RX, see www.izotope.com/en/products/rx.html. In RX the view includes the orange to yellowish color tones, being energy/sound by frequency. From this top-down view, the higher frequencies are to the right, the lower frequencies to the left. While the blue waveform is good for editing out or adding sections; cutting and pasting, the orange/yellow display is useful for visually locating sounds like a swallow, or rustles, and then removal of just that sound:

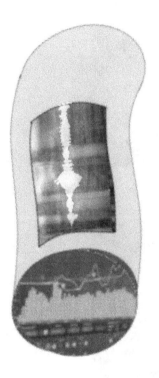

In the plugin software image below, representing the front-on view, I used the Waves TRACT equalization (EQ) plugin (used in Digital Audio Workstations). The real-time dynamic spectrum visualizer and controls will look something like this:

Similarly, here is an image of Izotope RX software, mentioned above:

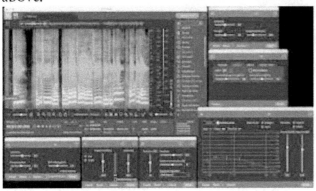

 **MIXING AND PANNING**

AUDIO MIXING is the process of taking any number of recorded **tracks** and blending them together, often to be exported as a stereo file. Each track is usually the recording of one instrument or vocal, and they probably have equalization (EQ), compression and other effects applied to make each of them sound right. As well some frequencies might be carved out (for stereo mixing), so as to let the sound of each track come through clearly, or panned left or right. Here are a number of tracks, as typically seen in Digital Audio Workstation software:

PANNING AUDIO you distribute sound across stereo channels, to create balance as if you were listening to a group of musicians in front of you. There are some established rules for where to pan different instruments to create this impression. Lead vocals are in the center, so split equally across the left and right channels.

Associated with panning left and right, is the positioning of sound forward or back in space. As depicted in the image below, different instruments are positioned back by the application of reverb, and an application of small amounts of echo for greater distances. So, piano audio is placed left, center, and right and, with a touch of reverb to set it behind vocals, guitars, sax, bass and kick. When listening, the music should then be 'imaged' (heard spatially) like this in stereo 3D space:

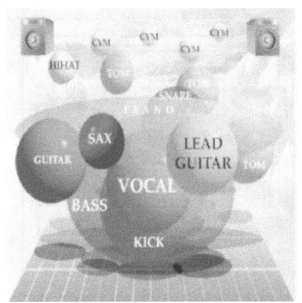

Image from "The Art Of Mixing" by David Gibson.

When listening to this mix in stereo, if the Hi-Fi speakers are positioned, typically, equally each side and in front of you, as listened to from the far point of an equilateral triangle, it will seem to your hearing as a soundstage, much like the picture above, with each instrument imaged in position.

Although a listener is mostly advised to sit in a centered position, tweeter seating, tweeter high frequency audio having the narrowest propagation of sound, is usually designed to spread the propagation to allow for reasonably wider listening positions. Also, there is the option of ortho-acoustic speakers, which reflect off the ceiling to create a canopy of sound as opposed to the more direct propagation of sound. In all cases the panning of panning/positioning of instruments in the original audio should be heard, with some application of acoustic treatment to duplicate echoing reflections.

If you play stereo music on a surround sound system it will default to play just from the left and right front speakers, like a normal hi-fi stereo system. This can be varied by selecting Multichannel stereo, creating the situation previously mentioned where the stereo is duplicated to all pairs of speaker. For music this may or may not sound better depending on the quality of your surround speakers, whether or not you have a subwoofer, the amount of room acoustic treatment, whether or not you have applies EQ, and the type of music you are listening to.

SURROUND SOUND

SURROUND SOUND is a different idea whereby the listener is in the middle of the stage, with the idea being for sound effects more than for music. For movies, the most basic cinematic surround format is 5.1. This requires 5 different full-range channel allocations: LFE bass, Left, Center, Right, Left-Surround, and Right-Surround speakers. The LFE, which is just for bass effects not bass music, goes to a subwoofer. But if you set your Bass management as LFE + Main it will auto adjust so that if you play stereo music the bass will goes to your sub as it is intended to. An (AVR) or Soundbar, renders the channels to the designated speakers, also upscaling and downscaling to handle differences between the channel count and the number of physical speakers you actually have. When music is included in the mix, they will still apply it equally to the left and right channels, and multiply the stereo to the other pairs of speakers with a tiny bit of echo so it will feel like the primary music is still coming from the front.

Panning and reverb placement is not needed or desirable for movie format surround sound, though, some bleeding/merging (referred to as Divergence) over channels is undertaken by the audio mixer. This is for movies.

The center speaker is dedicated to dialogue since it centers that dialog to the screen. To hear that, you will need an Audio Video Receiver (AVR) or Soundbar, for rendering the channels to the designated speakers, also upscaling and downscaling to handle differences between the channel count and the number of physical speakers you actually have.

If you are playing movies through a Hi-Fi system, the TV will downmix to stereo. Fine. But in the absence of a center speaker, the dialog may not stand out. I cater for this with a the McDSP SA-2 VST plugin in my JRiver Media player, to bring the dialog forward slightly (as discussed in the Lounge Listening chapter).

For video production, instead of an audio workstation, you will want a video editor that will handle video and audio well – meaning surround sound audio, VST plugin compatibility, and Dolby Atmos integration. In many cases you can connect a video editor and audio editor (Adobe Premier Pro and Adobe Audition or Final Cut Pro X and Logic Pro X), but I think a fully integrated editor would be a better choice if you are focused on video editing and wanting to edit for surround sound and Dolby Atmos. Have a look at Nuendo from Cubebase, which you can buy without a prescription, see https://www.steinberg.net/nuendo/. Also, see DaVinci Resolve Studio, https://www.blackmagicdesign.com/nz/products/davinciresolve/studio.

The mixing priority is to match the sound to what you are seeing. Here is a detailed article, by Waves Audio, on mixing for surround sound, see www.waves.com/mixing-in-surround-do-and-dont. Below we see how the 5.1 channels/speakers are referenced in the mixing process:

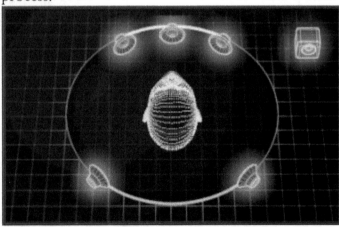

When surround sound is playing, it will depend where you are in the room, as to the different imaging of instruments you will experience. The experience is immersive, like walking around in the middle of the action. This understanding does lead to why center channels are used. They translate the dialog or song as if from the center of the screen, because a seen talking-head should always sound close to the center of the screen.

For more detail, have a look at this article from The Recording Academy's Producers & Engineers Wing, on Recommendations For Surround Sound Production, see https://www2.grammy.com/PDFs/Recording_Academy/Producers _And_Engineers/5_1_Rec.pdf.

DOLBY ATMOS

FOR OVERHEAD SOUND EFFECTS, if we consider stereo and surround sound, as we have discussed it, as a "bed of sound", Dolby adds "audio objects" at a higher level than that bed. The mixing requires the allocation of special effects to audio objects, that are moved in above-your-head 3D space. Multiple height speakers are needed to hear these overhead effect, four is a good benchmark. A Soundbar or AVR compatible with Dolby Atmos has a decoder to render out the objects to height speakers, in addition to surround sound channels. Here you can see the audio objects in the Dolby Atmos Renderer, placed to move in 3D space:

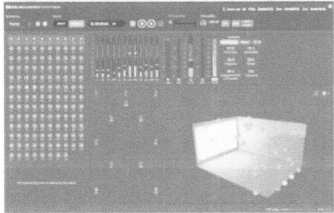

This image is of the Dolby Atmos Renderer used when developing Atmos content.

You can hear Dolby Atmos sound with a good pair of headphones and a computer or Smartphone. Without headphones, you do need height speakers or a soundbar that has some upward tilted speakers, which are used for the height channels. These details are covered later in this book.

SUMMARY

WE HAVE LOOKED AT how different sounds are produced within a frequency range and the resulting extended harmonics, the importance of certain frequencies, and how we visualize it. Mixing and panning processes are applied in order to create the sound as we hear it imaged on a soundstage in stereo, or as distributed for surround sound; perhaps with the addition of Dolby Atmos moving 3D sound effects.

CHAPTER 3
ROOM ACOUSTICS APPLICATIONS

"The acoustics seem to get louder", by Hugh Johns.

IF WE ARE LISTENING OR RECORDING IN A ROOM, reflections are an issue that muddy sound. There are three big issues: direct reflections, room modes, and room resonance. Not only do we miss out on emotionally connecting to the audio as intended, but we start to not enjoy being in the actual room, and seek another place. We simply lose attention.

Direct reflections are caused by produced sound bouncing off flat surfaces between the instrument or voice and the microphone, or between the speakers and listeners. The reflection may be off ceiling, floor, desk, computer screen, side wall, and the far away wall. These out-of-time-with-each-other reflections buzz-muddy the sound.

Room modes are static audio volume peaks in different places in your room, created at low frequencies from off flat surface reflections. These skew your bass sound as you move around your room, and make it indistinct like one boomy note. Modes are the hardest issue to solve, but if we don't, not only is our bass boomy and indistinct but the destructive harmonics buzz-muddy everything else at higher frequencies.

Room resonance, are also reflections off flat surfaces, but at mid to high frequencies, again skewing frequencies in different parts of the room, adding to the buzz-muddying, but when acoustically treated becomes a nice natural room resonance, a tone of airiness.

I show how different room acoustic treatments aim at resolving the issues created by **direct reflection, room modes, and room resonance** - both for lounge listening and recording spaces. And, we look in detail at how to make the different types of acoustic treatment, and apply them. When you get it right, audio in your room won't need to be as loud, and it will be distinct, clear, tonally pleasing, and visceral.

 ## DIRECT REFLECTIONS

IF YOU ARE IN A ROOM WITHOUT ACOUSTIC TREATMENT, the direct reflections will come from behind your speakers, your back wall, the side walls, the ceiling and the floor. These multiple sounds coming at you out of time are a problem.

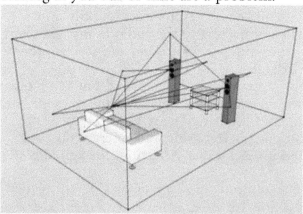

REF: *advancedacoustics-uk.com/pages/what-is-the-first-reflection-point*

The effect of these first reflections, because the delay is only milliseconds, but that is enough to starts to sound like a bad recording you make on your phone, fuzzy, boingy, not clear. Acoustic panels can fix this issue.

 REAR WALL REFLECTIONS

REAR WALL REFLECTIONS are particularly bad. Sound from speakers reflects off the back wall (the wall the speakers face), creating its own backwash volume peaks and drops/nulls. These back-wall reflected volume peaks occur for all frequencies at their 1/4 wavelength position and every quarter after that. There is a range of ever-changing frequencies from any music or movie; every note bouncing off the back wall onto itself, raising each frequency in volume at its 1/4 wavelength increments. These peaks and nulls muddy sound with numerous unnatural volume variations. Acoustic panels and diffusers can fix this issue.

The illustration below shows how just one sound you play reflects off the back wall, with the consequential frequency peaks and drops in volume. This means if there is a hard flat wall surface, wherever you stand in the room you hear a messed up sound, not the sound that actually comes from your speakers. The bright side is that the problem is less so in the front third of the room. In the proof of concept example below, a 39.34 Hz tone was played from front room speakers (apart from the first room 'Mode', explained from the link below), at 1/4 wavelength spaces of the 39Hz tone, at 3/4, 4/4, 5/4, 6/4 there are unnatural frequency related volume increases.

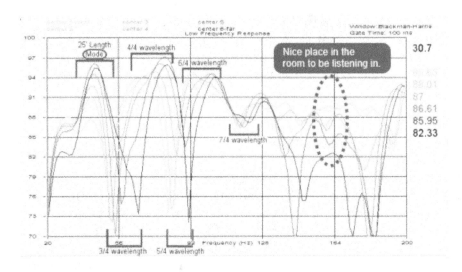

This data is described in detail at Real Traps, see realtraps.com/art_modes.htm. The degree that you can apply acoustic treatment to your back wall directly relates to the degree that you can hear clearly in different parts of the room. Direct reflection need to be reduced by Acoustic panels and diffusers can fix this issue. One exception is ortho-acoustic speakers, as discussed later.

You can try this out yourself, by using an Online Tone Generator, see www.szynalski.com/tone-generator/. Run the tone and walk between the speakers and the back wall, listening. The above tone was for a 25 foot room. If your room is half that size, try a higher frequency. I tested this in my lounge room with a 75 Hz tone, and could hear similar results walking between my speakers and the back of the room, with somewhere between 1/2 and 1/4 from the front wall being pretty even, as noted to be a nice place in the room to be listening in. And so, a good sweet spot for a listening and recording position is closer to the speakers (sometimes mentioned as 38% from the front wall).

 LISTENING SWEET SPOT

WE HAVE SEEN THAT THE BEST PLACE TO LISTEN in a room with traditional speakers is **about** 38% from your font wall. The speakers are preferably in a triangular position to you, placed as per recommended by your speaker manufacturer. This is the sweet spot:

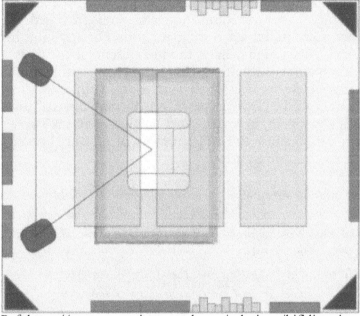

Ref: https://www.easy-noisecontrol.com/solutions/hifi-listening-room/

And, this is not very encouraging for people seated half way or further back in the room, where listeners (not audio editors) will often want to sit.

ROOM MODES

ROOM MODES ARE RATHER LIKE SEA WAVES crashing against a flat cliff face, smacking back into the next oncoming wave, which then rises higher than it was in the first place. Here is an example of water doing this, at the Bunbbury Backwash, see youtu.be/ 9BvrcutD2E. With sound, the higher the peak the higher the volume, and gaps between are lower in volume – higher and lower than the original sound. Unlike waves, sound is reflecting at the speed of sound, so the back-wash wave peak will be maintained in position.

The biggest issue in a room is caused by room modes under about 250Hz (about the sound of a car engine starting; the low part of a male voice – here the room modes are pressure based. We are exciting the room modes when we make sound of any kind in the room - even if it is a higher note, the mere excitement causes all room modes to come into effect. This is more so a problem if your room has concrete surfaces, such as in a basement, or concrete apartment walls. Gypsum partitioned internal walls* and ceilings are still a problem, however. And, smaller rooms are a bigger problem still, because the same number of pressured room modes are packed in more tightly.

> *Gypsum walls have their own resonating frequency (60-70 Hz), so will be very effective around that frequency at dissipating bass sound, and letting some lower bass frequencies seep through. The boomy noise, and low mumbling sound of voices you hear on the other side of the thin wall is what is dissipating out of the room. But while that mitigates the effects, the room modes are still substantially there in the room.

At amcoustics.com, see amcoustics.com/tools/amroc? l=470&w=480&h=245&re=DIN%2018041%20-%20Music, which is my own lounge room dimensions entered: length 470cm, width 480cm, and height 245cm. *You can enter your own room dimensions, which would be useful as you read through this.*

The first thing you will notice is a keyboard image with predicted room modes overlaid, up to 250 Hz, covering the bass area where room modes are the main issue:

When you move your mouse over the room mode red lines in Amroc, you can hear the frequency and also see a Room 3D graph window, showing where the room modes happen in your room. The light and dark areas (seen in Amroc as red and blue) are the axial modes, being the most important because they are caused by reflections off just two flat surfaces and so have the most energy. Here are the first ten axial room modes for my lounge:

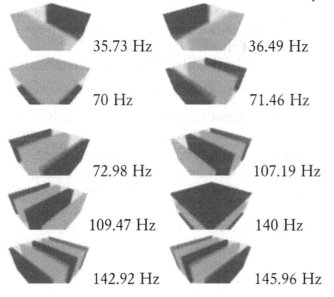

35.73 Hz   36.49 Hz

70 Hz   71.46 Hz

72.98 Hz   107.19 Hz

109.47 Hz   140 Hz

142.92 Hz   145.96 Hz

To get a good understanding of how sound works in your room, leave you mouse over one of the tones on the keyboard, look at the 3D image and walk into and through that spaces in your room (or do it with an Online Tone Generator, on you phone, see www.szynalski.com/tone-generator/. Play the first 6 to 10 of your own axial room modes frequencies, as seen in Amroc, one at a time, and step in and out of the theoretical position. You are learning to hear how your room messes with audio produced in it.

You will have noticed that all of the frequencies are louder in the corners, some along certain walls, some include modal peaks through the length of your room. In my case the 71.46 Hz and 72.98 Hz modes go through the middle of my room - so if I did have a listening seat on that line somewhere those two room modes would certainly affect the sound I hear from my speakers (unless I equalize them out, with room acoustics or other EQ methods).

The **35.73 Hz** and **36.49 Hz** cause room modes around the walls. Those are low bass and so will have significantly destructive harmonics in the bass region. Having furniture around the walls will help reduce the issues, but if you sit in a chair or couch against the wall the sound will be destructively affected. So, we can see that putting a bass trap against a wall will help reduce about half of them; in a corner, all of them.

Opening any doors and windows will also help reduce those room modes.

Some of the room modes around 60 - 70 Hz will go through partition walls. As mentioned, most internal gypsum walls have a resonant frequency somewhere around 70 Hz. When I played **70 Hz**, **71.46 Hz**, and **72.98 Hz**, with my ear against the internal gypsum walls of our lounge room, I can absolutely hear a hum in the wall - volume up, it shakes. That is a help. Also my two doors, if shut will hum at **72.98 Hz**, which I can hear, ear against the door. This means your walls and doors are acoustic sound absorbers at those frequencies. You can do your own experiments if you like.

PROBLEMS CAUSED BY ROOM MODES

THE ROOM MODE ACOUSTIC PROBLEMS in your room can be illustrated with a typical waterfall graph showing a sound sweep from 20 Hz to 3,000 Hz, and the watterfall-forward-dropping-away of sound:

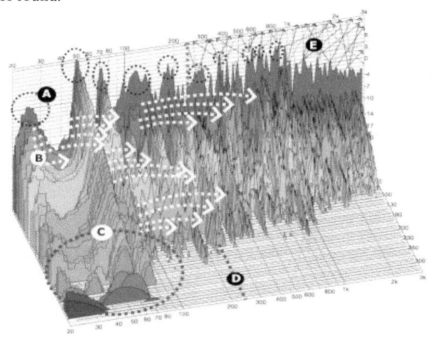

A sound sweep is a controlled volume sound played as it moves from very low bass of 20 Hz usually right up to a whistly 20,000 Hz, but, in this case up to 3,000 Hz. The dropping away of the front of the waterfall is showing how the sound fades away. The darker peaks are the fundamental sound that are made, inclusive of harmonics and room acoustic interference. In the image I have numbered the different interferences A-E. We see:

(A) Room mode peaks.

(B) Harmonics, from fundamental sounds, and room modes. The

27

harmonics made from room modes destructively influence the tone higher up. See this by John Heisz Speaker and Audio Projects, youtu.be/vr3kmMRWwBw.

(C) Fading away of the sound (decay time) of the bass level room modes takes longer than the fading away of other sound frequencies because of bass sustain. This causes modal ringing, which in turn causes a one-note bass as explained by Ethan Winer of RealTraps, see youtu.be/aHkAFSZmMk4. Ethan goes on to show how the one-note bass is significantly resolved with acoustic bass traps. A similar result can be achieved with EQ as explained by Bob Katz, see youtu.be/ ym-CazJa M). In Bob's video you can listen to the one-note low hum bass before he applies EQ to bring down the offending 124 Hz mode; after which you can hear clearer distinction in the bass. It is, however, always better to resolve the issue with acoustic treatment because that is preventing the harmonics of the modal ringing, whereas EQ will address the mode only.

(D) The Schroeder frequency (somewhere around 250 Hz) is a transition zone, after which room modes matter less, because after that point the room echoes fuse together into a single coherent sound, thereby having a more natural overall tonal sound. However, as we see in (E) we still want to avoid that tone becoming boingy.

(E) Resonance (commonly known as echo) above the Schroeder frequency, is caused by sound bouncing off different flat surfaces in a small room and then arriving at your ears out of time. It can destructively influence the resonant tone above the Schroeder frequency, causing a metallic boingy sound, like a poor recording of your own voice – and clarity suffers, see this explanation by Geo Martin, youtu.be/6Q0joik6E74?t=477.

There is a "critical distance" where beyond it the room resonance dominates, and it is hard to hear clearly, see youtu.be/6Q0joik6E74?t=1140. Bigger rooms with acoustic room treatment extend the critical listening distance more so out into the room.

True 'pleasant' reverberation of produced music is when relatively even audio reflections fuse together into a coherent sound, bringing spaciousness to the sound - a good thing - increasing the intensity and richness of tone, while at the same time there is the right measure of direct sound. Resonance is not bad in itself, because we expect it to provide a sense of space, but when there are too many flat hard surfaces the metallic 'boingy' sound becomes destructive.

THE SOLUTION? Discussed with build instructions below, is to build bass traps, perhaps **corner Supertraps** [about 400mm deep], placed as you are able, in corners where walls and walls, and walls and ceilings meet, **and/or hybrid membrane bass traps**. As well, add 1-4" **broadband panels** to reduce reflective echo, placed at direct reflection points, including behind speakers and, if possible, the ceiling above a listening or recording position. As well, add **diffusion** to scatter sound, useful for higher frequencies and sound off the back wall. As well, floor carpet, rugs and furniture, and people all help.

DISCUSSED BELOW, one should have more acoustic treatment for a recording space compared to a media listening space. This is because in the listening/media playing space, you can use, (a) surround sound and subwoofers, which in themselves help reduce acoustic issues, and (b) and you can use headphones. Whereas, in a recording space you are capturing sound that includes any destructive room reverberation - DeReverb and EQ applied after recording can help, but not eliminate the issues.

 **LOUNGES AND ACOUSTIC TREATMENT**

IN A LOUNGE YOU WILL BE LIMITED as to how much acoustic treatment you can apply.

One of our lounge walls has two doors which can be left open. Open doors are an excellent way to diffuse the wave energy, like this water gap, by Alexander Mathematics & Physics Tutoring, youtu.be/6cNrh92kcRU. Here is my lounge:

From left to right: the hanging Turkish carpet has a polyester acoustic panel and pegboard behind it. The doors if left open reduce potential direct reflection off the wall, between the two doors I have two stacked diaphragmatic bass traps (with inbuilt shelving as a decor preference). One the right side of the wall, there are hung two diffusers, the picture, between them is off the wall slightly with acoustic polyester behind it. In the corner, tucked beside the couch, there is another diaphragmatic bass trap which serves as a side table. The couches also help. Each of these acoustic treatments are explained in details below.

I use a 4K TV as an extended computer monitor, switching the AV input to TV for watching TV or Netflix. In extended monitor mode, the 4K allows for excellent textual detail. And, with the background black, it is not too imposing. In the pic below I am working on a document with Tidal music playing to one side. Equally, I can be browsing the Internet. Audio is from the computer to a DAC, on-connected to a Denon AVR, which directs audio to speakers or my headphones. I plan to add a Schiit Freya+ pre-amp between the DAC and Denon to enhance analog-like holographic listening:

I also use this setup for watching TV, arrived at because my wife is more of a reader and does not want a TV as a central feature in our lounge. I might watch YouTube, leaning back in my desk chair (with headphones, while my wife reads), or, more so for movies or YouTube, stretch back in the chair or lie out on the couch (headphones or not), and on the occasions when we watch a movie together, I slide the couch around to face the TV:

You can see one of my ortho-acoustic speakers at the side. Normally, when I am at the desk, you would think this speaker is too close, but being upward-firing it is super versatile in terms of placement. The AVR can auto-adjust volume based on the listening position. There is also a center speaker below the TV. The side cabinet stacks a Denon AVR, a router, and a subwoofer.

Audio is from the TV via HDMI ARC to the Denon AVR, which directs audio (including Atmos) to a 5.1.2 speaker setup, and to headphones.

Above, you can see my stack of diaphragmatic bass traps in the corner, inclusive of shelving. And, behind the curtain, I have a 4 inch thick broadband bass trap, as seen here:

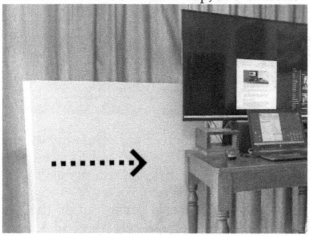

A SMALL RECORDING SPACE

I HAVE A RECORDING SPACE DOWNSTAIRS (L640 x W263 x H240), which at the moment I am using for online teaching, but I also use it for off-line recording. The room's dimension is not acoustically wonderful, being more narrow than I would prefer, but I stack in acoustic treatment to make it work well. Here are the room modes I needed to deal with, see amcoustics.com/tools/amroc?l=640&w=263&h=240&re=DIN %2018041%20-%20Speech:

The room is downstairs in our house and, being able to dedicate the space for recording, I was able to be far more generous with acoustic treatment, and indeed it is more important for recording. The same amount of acoustic treatment was used for the small space as you would use in a larger room. I put a large braodband acoustic panel 5" (127mm) thick above the microphone recording position, and similarly on both sides, and also against the back wall. For these broadband panels, I used common household glass wool. I also put corner Superchunks where the wall and ceilings meet. Behind the chair, and to one side, and in front of me, and then added 280mm deep membrane bass traps. Lastly, I added QRD folded diffusers, to the back wall and to one side wall so that the room did not become overly "dead" from the absorption. These acoustic treatments are explained below.

At first sight you might think it is overkill to have so much treatment. Not so. It is a small space, and so the same amount of room modes are crammed in there. To release the energy you need the same absorption capacity as if it was a large room:

I normally stand or sit on a bar stool when recording. The extended screen above the computer is just above my mouth height, and I partly fold down the laptop screen so the membrane bass trap, on which the extended screen is supported absorbs a good amount of my direct voice. Likewise, I have placed broadband absorption behind and beside the microphone.

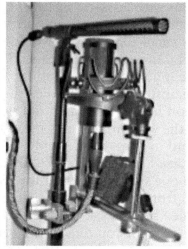

A SEGUE TO CONSIDER MY MICROPHONE SETUP:

I have a dual microphone setup (ribbon and shotgun condenser). I do this to get the tonal benefit of the ribbon and the accuracy of the shotgun condenser. The ribbon mic is a <u>Shure KSM 313</u> [which is figure 8 bi-directional] (connected through a Grace M101 preamp to my digital interface), the shotgun condenser mic, a <u>Synco Mic-D2</u> [which is hyper-cardioid polar pattern] is over the top of the ribbon mic. When online, the distance of the microphones is about a foot from my mouth, which is just out of camera view - for off-line recording I reposition the microphone setup at a finger-spread distance from the microphones.

For two mics, like the above, the pickups of the two mics need to be lined up. It is worth experimenting with test placement, then making a stereo recording, zooming in to look at the alignment of the two tracks. Below, I hit two pens together for a sharp sound, then zoomed in for a close look at the waveform. Initially it was not aligned. I then moved the shotgun mic forward over the ribbon mic and retested until I got it aligned, as can be seen in the overlaid image compared to the image behind where the timing is out. You can also check for phase alignment (that the waves of both tracks and going up and down together - if not, flick the phase switch on one of the mics, or in your recording software:

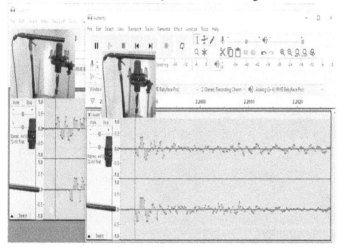

If you are recording stereo from two mics into a Digital Audio Workstation you can also align the two tracks in post, then merge them into one track, then edit. I have detailed this merging in the chapter, 'Clean Audio'. You can handle the alignment with a tool like the Azimuth module of Izotope RX Advanced, see www.izotope.com/en/products/rx/features/azimuth.html. Other plugins do the same thing. To consider options, search "Audio editing and alignment VST plugins".

I have provided some information about the camera setup and camera alternatives in the chapter, 'Recording spaces and room acoustics'.

NOW LET'S SEGUE BACK TO THE DIFFERENT KINDS OF ACOUSTIC TREATMENT:

BASS TRAP: REFLECTION FILTER

A REFLECTION FILTER, like the sE Elecronics RF SPACE, is another way you can acoustically insulate a microphone from room echo, for example, where there are limits on how much acoustic treatment you can apply. I have used one of these, as well as acoustic treatment for recording with a ribbon mic, to good effect. However, a reflection filter will add a little bit of its own resonance. Therefore, it is a compromise, such as if you want to record while looking toward a window. It is better to rely on room acoustics if you can.

 BASS TRAP: SUPERCHUNK

A SUPERCHUNK IS A BASS TRAP commonly designed for recording spaces. They are a great cost-effective and easy-to-build solution. Indeed, superchunks are the most common bass trap you will see people building on YouTube - either that or very thick absorption panels. Household insulation, bought at your local home depot, is the best material for building superchunk bass traps, and the best place to put them is in the corners of your room (also, along the edge where your walls meet your ceiling, as I have done (see below).

A typical depth for superchunk bass trap is 13" (33.02 cm), so the face would be 24" (60.96 cm) across, the sides from the corner 17" (43.18 cm). You can make triangular frames, or attach triangular brackets to your walls, then fill in the space with acoustic insulation, then cover the face with breathable material to blend with your decor, something like this:

Here are some instructions by Sandi on the hifivision.com forum, for building a movable Superchunk, see www.hifivision.com/threads/corner-bass-traps-diy-caution-picture-load.54553/.

And, here is a links to some building instructions for fixing a superchunk to the wall, by jlafrenz in the Emotiva Lounge forum, see emotivalounge.proboards.com/thread/54838/superchunk-corner-bass-traps-tutorial.

Sometimes you will see these built with heavier insulation, but household wall insulation is actually the most effective – the softer density working best for the depth.

Here is my own custom Supertrap for along the wall-ceiling edge in my recording space. Here it is hung up in the wall-ceiling juncture (there is an eye hook screwed into the beading, with cord threaded through to secure it in place. I add these, one after the other:

These are 52 cm abreast from longest corner to longest corner, and 40 cm from the front face to the wall-ceiling corner. I designed them like this because they are little less obtrusive for a ceiling, and I can butt up other broadband absorbers to them, plus, the design gets around the ceiling-wall beading.

Building instructions:

For the wooden frame I used 25 mm fence palings which are about 150 cm wide, first ripping the palings to exactly 150 cm wide, so they were the same. The fence palings were then cut with a drop saw into 52 cm lengths (cutting at 45 degree angles). I then used the drop saw to cut approximately the inside angled cuts, and the inside length, then put them in a vice and hand sawed to get the finished cut-out:

Secondly, I glued two together to make frame pieces:

Thirdly, I used three frame pieces and joined them up using gypsum corner beading cut to 120 cm, reinforcing them on the more pointy extended edges with wooden strips, also ripped from my fence palings. These are pretty light, which is what I wanted, so they would be easy to handle. I used button head screws when fixing the corner beading to wood:

I filled them with semi-rigid insulation, and stapled on some breathable material. In the picture above you can see some white material tape. I used it to hide the stapled material on the end. This only needs to be done for an exposed end – one could make a nicer looking end panel for any exposed ends. To hang them, I put some eye screws on the inside edge, and the same into the wall-ceiling beading. Some line was then used to haul them up, and that was secured on an improvised small cleat made by two button head screws (see above).

 BASS TRAP: HYBRID MEMBRANE (DIAPHRAGMATIC)

NOW, FOR THE MOST EFFICIENT BASS TRAP I can find, and which can be applied not just to recording spaces, but to lounges, home theaters and listening spaces.

At first glance it may look hard to build, but if you use kitset bookshelf cabinets as the first building block, it is a job anyone could do, with a few instructions that I provide below.

It uses a rubbery membrane to act as a diaphragm, most commonly mass loaded vinyl (MLV); the MLV membrane vibrates out low pressure bass energy. Something similar is the soffit bass trap design by GIK Acoustics - see their "Range Limiter" model, https://gikacoustics.net/product/gik-acoustics-soffit-bass-trap/. The design pairs acoustic material behind the MLV to broaden the effect, with the design below including additional acoustic material on the front to double-up the design so it is also a broadband absorber.

Here, Bilou, at Gearspace, see gearspace.com/board/showpost.php?p=8478913&postcount=455, has made a Bass Trap of this type. It is 280 mm thick; the thickness of a bookcase (Ikea' Billy bookcase, in his example), and so works well for the decor. Bilou, he explains, made four of these for a client: two for the back of the room, and one for each side of the main listening position:

From the forum link you can see Bilou's "before" and "after" waterfall graphs from Room EQ Wizard (REW), showing the room's modal peaks being pushed down about 6 dB, and the reverberation time of the low frequencies dropping off sooner and more evenly, somewhere around 80 m/s (0.8 seconds), which is as good as it gets for bass trap effectiveness.

The building method for the Hybrid membrane

(diaphragmatic) bass traps is like this:

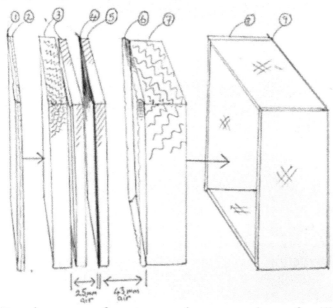

Yes, there are a few steps and you may be unfamiliar working with Mass Loaded Vinyl, but if you make a number of modular ones even with integrated bookcases, you will never be sorry.

(9) For the back of the cabinet, use 12 mm MDF from home depot. If attaching to the back of a bookcase that you are putting together, glue the sides that will form the enclosed box (not just relying on nails or screws). This MDF backing you attach, in the case of an Ikea bookcase, would be replacing their very thin backing. An Ikea Billy bookcase is 800 mm tall. If you are looking at other bookcases, to avoid the cabinet vibrating, keep the bass trap size within 800mm - 1000mm tall. If you want to make the bass trap part bigger, build a divider into it. Any bookcase, if 280 mm deep or more can be used. If deeper than 280mm it will simply reach down to lower bass frequencies, which is also fine. Using a bookcase like this has the advantage of it being good for the decor and it becomes multi-purpose as a bookcase/storage

space. You could similarly utilize any sideboard, cupboard cabinet, or storage cubical system.

(8) If you make a separate box, or make your own bookcase design, like I did, a good option is 16 or 18 mm MDF for the sides, which you can get from your home depot. A home depot will usually cut it in strips for you (Then you drop-saw the lengths you want, or ask them to do that too).

(7) Against the backing, use 150 mm Glasswool/fiberglass wall batts (which will have a density of about 20kg/m3). This is household insulation for walls (sometimes listed as R2.5). I combined this with acoustic polyester panels for an outer layer, so I did not need the next 6th step. Note: 150 mm insulation, then 43mm air gap = 193mm gap behind the MLV to the back of cabinet, which is targeting 52Hz, nice and low to start from, and a common area of room modes. The insulation broadens that target to reach down to as low as 40Hz.

(6) Use plastic fly screen to hold the insulation in place. You can staple it to the inside of the box.

(5) Fix wooden batten around the inside, which could be made from pine, perhaps 19 mm thick cut or bought to a width of 30-40 mm. I saved money by ripping them from fence palings. Screw and glue the wooden battens to the inside of the box, making sure it is 100% sealed with acoustic sealant. The air between the MLV and the back of the box must be completely sealed so the air acts as a pressure spring. I used polyurethane adhesive sealant to ensure the seal.

(4) Use a 3 mm Mass Loaded Vinyl (MLV) layer, which is 5kg/m2 (1 lb/sq ft). See if you can get it from a builders supply company. Sometimes the material is used to line walls for sound insulation. Get enough for four modules, so at least 4 meters in length. Apply some polyurethane adhesive sealant to where the MLV will lie against the pine frame, then it will be stapled to the frame. The idea is for the MLV to be limp but not saggy, so secure one side,

then turn the cabinet around on its side so the MLV hangs, and then secure the other edge, and repeat for the other two sides. I actually supported it with temporary polyester panel wedges, while I fixed the first three edges.

(5) Add an on-top batten frame on top of the MLV edges (pine 18 mm or a bit thicker), screwed down on the MLV and with more polyurethane adhesive sealant at the joins. That will further secure the MLV and provide a raised edge for the next layer - so the outer insulation will not ever touch the MLV. Staple some flyscreen to that on-top frame. Now, add 50mm of Rockwool 90kg/m3, or equivalent. Try and get this from a building or acoustic supplier. The 90kg/m3 is the amount of density, meaning the amount of compression applied when manufactured. The density can be slightly different, anything between 80 -100 kg/m3. After cutting it to fit, remove it so you can do the next step.

(6) Make a frame from 18 mm pine trim to support the final material cover. A way to snug fit it is to cut it to fit snug with temporary bits of double layer facing material placed on two sided. This way the final covered frame will be a snug fit so you do not need to worry about velcroing in the frame. You will need to cut out a little of the 500 mm rockwool, so it fits - do that and replace the Rockwool. The frame will hold the Rockwool in place.

(7) Secure some facing material, which is a little breathable, to cover the frame, not shiny. Polyester or upholstery material are good choices. Because it is a base trap it does not need to be as breathable as Hessian. Wrap the material around the frame and staple it. Lastly, snug fit it into place.

Here are some build pics, this time numbered in order of construction:

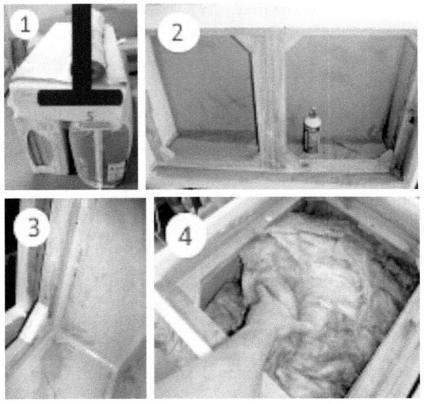

1. Materials ready to go.
2. Cabinet made, wooden batten support for mass loaded vinyl layer completed.
3. Internal joins sealed.
4. Glass wool in pace (150mm).

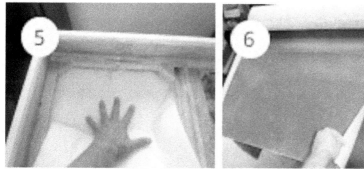

5. My method of adding a layer of polyester acoustic panel to maintain the air gap. Alternatively the softer glass wool can be held in place with stapled plastic fly screen mesh.
6. Cut Mass loaded vinyl (MLV) to size.

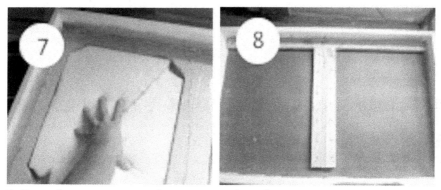

7. I used temporary polyester panel wedges as support when securing the first 3 sides of the MLV, so it does not sag as I fix it in place. These temporary wedges I cut in two so they can be easily pulled out before the final edge of MDF is secured. Also, polyurethane adhesive sealant is smeared onto wooden frame support, and to the back of MLV where it will join.
8. With MLV in place, some of the top layer of wood framing is screwed down, holding the MLV in place. It is hardly sagging because of the temporary support behind it. The MLV should not be stretched tight – limp, but not sagging.

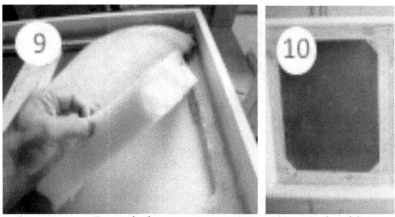

9. In my case, I needed to remove my temporary backing support before finishing the securing of the wooden frame.

10. MLV fully secured. A tiny bit of sagging is there, no issue – being sealed, if you push in one corner the other corner will bulge out. You could add some flyscreen over the frame so bits of insulation do not fall over time onto the MLV.

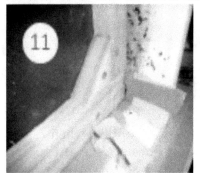

11. Doubled-up facing material is temporarily placed on two sides as a spacer when building the outside frame, so the facing frame will end up as a snug fit.

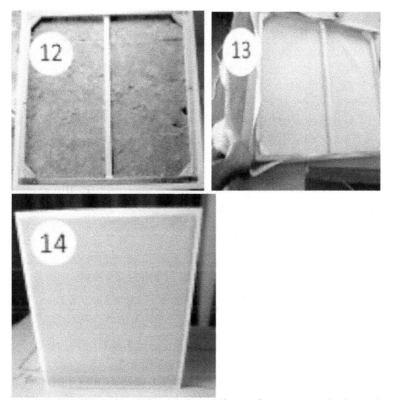

12, 13, 14 Here is a YouTube video of me completing the facing frame process, see youtu.be/wXlhGeI6hRA. For 12, a top layer of Rockwool is placed over the wooden frame that holds down the MLV. Then a final facing top frame is made to snug fit. The frame is made with glue that was labeled as suitable for framing, because it expands into the joint as it dries. For 13, the fabric is then stapled to the facing frame. You staple the four center positions, then work your way out to the corners. For14, the completed covered top frame is now flipped and snug fitted, no need for Velcro or spacers. Da-da! A finished, hybrid diaphragmatic and broadband absorber bass trap. Yes, that last part is hard to explain, which is why I included the YouTube link.

THESE MEMBRANE (DIAPHRAGMATIC) BASS TRAPS, on their own, without being covered with the front level of absorption panel, is meant as a second stage of acoustic treatment, a strategy more often pursued for a professional recording studio setup. The internal depth of the cabinet from the inside back to the MLV layer targets specific frequencies, so if you measure your room after applying acoustic treatment, you will see some peaks you might still like to bring down/target. The additional front layer of broadband absorber is not needed in this case. Here is the calculation, for the peak resonant frequency of the membrane (diaphragmatic) bass trap build above, but not counting the effect of the outside broadband treatment:

fo = fundamental frequency
fo = 510/square root of MxD
fo = 510/square root of 5x19.3 (5 means 5Kg/m2, 19.3 is the 19.3cm depth behind the 5Kg/m2 mass loaded vinyl)
fo = 510/square root of 96.5
fo = 510/9.82344135219
fo = 51.9Hz.

Therefore, a cabinet depth of 19.3cm enclosed with 5Kg/m2 mass loaded vinyl will reverberate 51.9Hz (range 37Hz – 76Hz) energy out of your room. I made the same calculations to arrive at in the bass trap build described above*, the range being what is more widely covered because of the outside broadband treatment:

[using MLV 5kg/m2]
1.Targeting 40Hz (range 30Hz - 60Hz) Internal depth of box 33cm.
2.Targeting 50Hz (range 38Hz - 75Hz) Internal depth 20.8cm.
3. * Targeting 52Hz (range 37Hz - 76Hz) Internal depth 19.3cm.
4. Targeting 63Hz (range 47Hz - 95Hz) Internal depth 13cm.
5. Targeting 80Hz (range 60Hz - 120Hz) Internal depth 8cm, *or if MLV 2.5kg/m2, then depth 16cm.*
6. Targeting 100Hz (range 75Hz - 150Hz) Internal depth 5cm, *or if MLV 2.5kg/m2, then depth 10cm.*

[Using MLV 2.5kg/m2] *Targeting 125Hz (range 94Hz - 187Hz) then depth 6.72cm.*

Always make make 4 or more, to achieve a noticeable effect.

 ## BASS TRAP: THICK BROADBAND PANELS

LIKE SUPERCHUNKS, THE THICKER BROADBAND PANELS ARE BETTER, so, doubling or tripling up some R.2.2 insulation is ideal. Here is an example of a thick broadband panel:

REF: *www.acousticsinsider.com/blog/best-insulation-material-diy-acoustic-absorbers.*

For anything over four inches thick you do not need specific acoustic insulation; use insulation from your local home depot, which is more effective, such as:

*Earthwool /glasswool household **wall** batt insulation, such as available from your local home depot.*

R2.2, *90mm thick, 14kg/m3, airflow resistivity mks rayls/m (Pa.s/m2) 6,000, is best to use for 90mm x2 or thicker R2.4, 90mm thick, 15kg/m3, airflow resistivity mks rayls/m (Pa.s/m2) 6,800.*

NOTE: rayls/m works out the same as Pa.s/m. The R2.4 is a thermal value, but it shows you what to buy, because the other data probably won't be on the labeling.

BROADBAND PANELS

BEYOND ROOM MODES, THERE IS STILL THE ISSUE OF THE DIRECT REFLECTION POINTS, as shown at the top of this chapter. Reflected sound causes a metallic boingy sound that suffocates tonal clarity. Here again we see where the typical first reflection points:

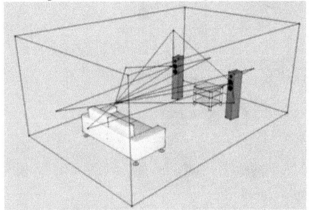

REF: _advancedacoustics-uk.com/pages/what-is-the-first-reflection-point_

A good solution for this is thinner broadband panels, 50 mm, 2" thick, traditional acoustic panels. Being less intrusive, these can also be mounted on a ceiling direct reflection position. As explained below, in my lounge, on the wall only, I used an acoustic panels behind a hanging Turkish rugs. In my recording space I have a thicker 5" panel above the recording position. Two inches thick is fine for wall panels and ceiling panels in a lounge:

RE: Image from site referenced below

Here is a good link for these DIY broadband panels at aciousticsfreedom.com, see <u>acousticsfreq.com/how-to-build-your-own-acoustic-panels/</u>

If they are 2" (50mm) thick, and if there is a small air gap between the panel and flat surface, they will cover 500 Hz - 4 kHz at a direct 90 degree angle. The important frequencies in non-tonal (Western) languages, illustrated by the diagram below:

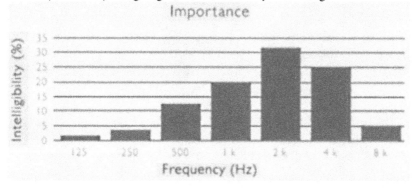

I would suggest using acoustic polyester panels for a ceiling panels, because they are light, rigid enough, and you won't have to address any concerns of having glass fiber above you. Cover with Hessian or other breathable fabric.

One absorbent facing material is wall rugs, like our Turkish rug, below. I have stacked a polyester acoustic panel behind it, then pegboard with some alternate size holes added. The optional pegboard further lowers the frequency range:

You can also add insulation to the back of pictures and extend the picture out from the wall a bit. It is extremely easy to do this because you already have the frame of the painting to slot the treatment into. Yes, there will be reflection from the face of the painting at higher frequencies, but it is still an improvement:

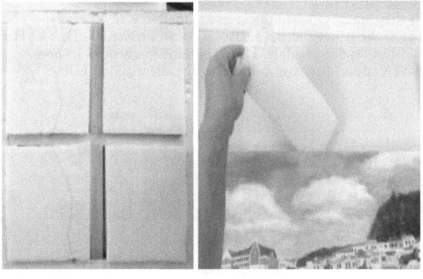

Materials for Broadband absorbers:

NOTE: John Sayers, a well know acoustician, recommends 32kg/m3 as an ideal density for broadband absorbers, commonly 18,000 mks rayls/m (which is air flow resistivity).

Any of these or similar would work well:

1. Owen's corning 703: 2" (50mm) thick, 48kg/m3, 18,000 mks rayls/m).

2. Roxul Safe'n'Sound: 3" (75mm) thick, 40kg/m3, 12,000 – 16,000 mks rayls/m.

3. CSR Bradford (Australia) Supertel 32kg/m3, 18,200 mks Rayls/m.

Fiberglass insulation can be covered with <u>Dacron</u>, if you like, to prevent glass particle pollution.

4. Polyester acoustic panels, as suggested for ceilings, can also be used: 50 mm thick, density 32-60Kg/m3, Airflow Resistivity 8,000-21,000 mks rayls/m. (https://gearspace.com/board/showpost.php?p=14249018&postcount=177)

Frames are often built with pine wood.

Broadband panels are not the only treatment we can put at direct reflection points – we can and should also use diffusion.

 **DIFFUSION**

IN THE PICTURES BELOW, YOU CAN ALSO SEE SOME SLATTED PANELS. These are called QRD diffusers, in this case QRD folded diffusers. Diffusers scatter sound while maintaining natural sound energy in a room. If we only have bass traps and absorbers in a room, such as we have looked at so far, it will start to sound unnaturally dead, the bass traps and panels sucking out all the energy. We have noted that above the Schroeder frequency (above 250Hz), reverberation of reflections dominate (much like sonar reflections). So, if we don't do anything about the remaining reflective flat surfaces, we will also end up with a 'hollow boingy'

sound, to a lesser extent like sound in a bathroom.

We need some diffusion, and we need more or less depending on the kind of recording or listening we are doing, less for talk, drama or pop music, more for serious classical music. And, we need more diffusion in larger rooms to manage the larger flat wall surface reflections.

Here is a good explanation by GIK Acoustics on the concept of diffusion (along with some of their diffusers, see youtu.be/1MtJOGXVZ1w. One of the issues discussed in the video is that diffusers need to be four feet from the listener or microphone to be effective. Actually, this depends on the design, that being the reason I recommend QRD diffusers for small rooms - they are harder to build, but you can be closer to them than the other types.

It is useful to see a perfectly set up room and how much diffusion is in it, and what it looks like measured. Here is a video by Doug Ferrara, at RealTraps, doing just that, see youtu.be/dB8H0HFMylo. In short, you cannot have too many bass traps, broadband panels work well on the ceiling, and diffusers also work well at direct reflection points on the walls, better than broadband panels (because of the energy factor). Here is the measured room from the video before any treatment:

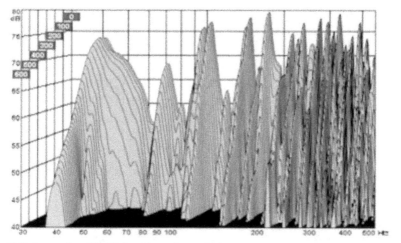

RE: Image from YouTube video referenced above.

Here is the room after treatment, which is as good as you can expect:

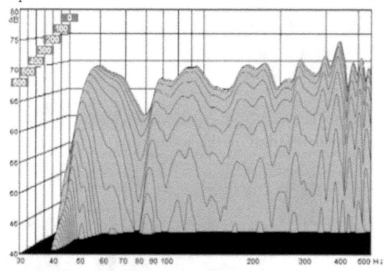

RE: Image from YouTube video referenced above.

The way to measure reverberation is usually by RT60, being the time it takes for 60dB of sound to fade away in 60 seconds. In the image above, from 100Hz upward the ridge line down to the ground is very even showing the RT60 as being around 200 milliseconds (0.2 seconds). The degree of peaks and valley ranges at worst by 15dB, mostly with 5dB variation. So, that is as good as any room is going to look when measured. Here is a measurement of my own recording space, measured using REW, see www.roomeqwizard.com, and a measurement microphone:

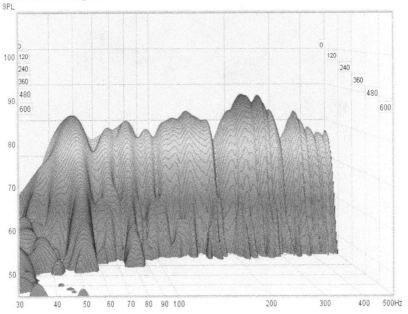

Again, here is the treatment in this room:

LET'S LOOK AT HOW TO BUILD DIFFERENT TYPES OF DIFFUSERS:

Diffusion: Slats over superchunks or broadband absorbers

You can make a hybrid diffuser by adding slats to a corner superchunk, something like this one from GIK acoustics:

REF: *Image source unknown*

Adding slats to corner superchunks, or any broadband acoustic panel is a relatively easy option. Whatever kind of corner superchunk you have, this requires making a front frame, and adding slats over the top, and fixing it to the face. You can see quite a lot of these with nice designs that you can buy.

Here John Brandt provides some design guidelines, see gearspace.com/board/showpost.php?p=5765376&postcount=25. And John Brandt's designs, see jhbrandt.net/resources/, under, "broadband with slats" designs) They are 51 mm x 19 mm Slats with gaps of 19 mm. Increasing the gap between slats to 25 mm is useful for a small room to maximize absorption of the superchunk behind.

Diffusion: Polycyndrical

A POLYCYNDRICAL DIFFUSER is achieved with a convex shape. There is one way to make them that is easy:

Use rounded floor tile edge trim, as above (available at home depots), securing edge trim to the wall or back of a door. Then, bend a piece of thin ply into it, over-sized to get an arc (dampening the ply and leaving it overnight will make it easy to bend). If you make more than one, spread them out rather than putting them side by side, perhaps interspersed with broadband panels or QRD diffusers. If you have limited space, make the convex curve less so, so as not to intrude into the room too much. If you like, add material – or a canvass painting, to the ply and paint the edge trim appropriately. You could even add LED lights shining from inside for an effect. You do need to be 4 feet (1.2m) from them.

What would be a very cheap, effective enough, and easy DIY acoustic setup, for a recording space be like?

1. Fill your room corners with Corner Superchunks, up to the ceiling. Use stacked Earthwool/glasswool household wall insulation batts, probably labeled R2.2, as available from your local home depot, similarly for (2). Keep the Earthwool in its plastic roll, cut holes in the plastic, without it breaking, then cover with an old blanket.

2. Make some very thick floor standing Bass Trap Broadband Panels, at cabinet height, to stand on the floor across the wall space, interspersed with a few actual cabinets and/or book cases* to be practical, all at the same height. You can probably buy a number of cabinet shells (without the doors, from your local home depot. After placing Earthwool inside the cabinet shells, cover with Hessian, stapling it at the back or folding it around the outside of the Earthwool slab. * Get a couple of cupboard doors, so a few empty cabinets can be useful storage cabinets). Just make sure you do not have any solid door cabinets at direct reflection points.

3. Standing on the thick Bass Trap Broadband Panels and any same height cabinets and/or book cases, intersperse thin plywood Polycyndrical Diffusers (as shown above) with Broadband panels. Design both so they do not stick out as much as the very thick floor standing Bass Trap Broadband Panels, so there is some useful shelf space. It is better if they reach to the ceiling. If you are renting, use backing plates for the Polycyndrical Diffusers rather than securing them to the wall, or instead of the plywood Polycyndrical Diffusers, use artificial plants and/or leaf or hedge tiles.

4. Make a large, or a couple of large Broadband Panels and hang them from the ceiling above the recording position. Use Polyester acoustic panels for these. Hang them using eye hooks and cord that you can tie securely with cleat hooks. Polyester acoustic panels are white, so you do not need to cover them or make frames for them. If you are renting, secure two cords from the top of two wardrobes, and/or curtain tracks, and balance your polyester acoustic panels on the cords.

Diffusion: QRD

QUADRATIC RESIDUE DIFFUSERS (QRD) have a series of wells that reflect sound back in to a room in a fan like array. The QRD design benefits from a lobing effect caused by the mutual interference of the scattered out of time wavelets reflecting from the wells. This means you can be closer to them than other diffusers. Here we see the lobing effect:

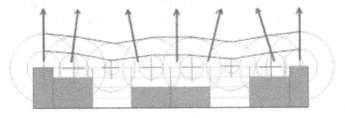

REF: Image from https://www.subwoofer-builder.com/qrd.htm

The best design I can find is the **folded diffuser design** by ThingMan (ReadScapes), which minimizes how far into the room they extend. For the fwd36 (shown below), you can sit 0.7m (2.3') from them with a sweet spot at 1.5m (5'), and for the fwd40, you can sit 0.8m (2.6') from them with a sweet spot at 1.7m (3.3'). Here is the cross section:

REF: Image from https://www.subwoofer-builder.com/qrd.htm

See all the details, and building instructions, by ThingMan at ReadScapes, see dngmns.home.xs4all.nl/fwd_uk.html.
For diffusers to work most effectively, place them together on your back wall or other places that are direct reflection points:

I have built a number of these, 22, actually. Well, they are a little tricky to make, but it can be done if you have a few instructions, and you will be very happy to have them for their effectiveness.

You could get your home depot or supplier to pre-cut the strips that you need. Otherwise you will need to buy sheets and cut them down (which you can do yourself with a rip saw). Plan on building at least eight, so as to be effective in a room. You do need a drop saw. If you do the smaller one (**fwd36**) a 10 inch drop saw blade will do, If one size up, the **fwd40**, you will need a 12 inch blade. The drop saw is needed for trimming the ends absolutely flat before gluing on the end-plates. Alternatively, don't bother with the end-plates.

A variation to the building instructions from the link above is to (a) use masking tape, rather than screws, to hold the pieces in place while the glue is drying. That will work if your use external PVC glue because it sets more quickly than the cheaper internal PVC glue. My son-in-law is a professional cabinetmaker, and their factory joins all MDF furniture like that because screws, by the way, push alignment off by 0.5mm, even when you pre-drill. And a second variation to the instructions in the link above, is to (b) paint them as you build, in two sections (see section shading below). Just don't paint where you will glue the two sections together. If you do not paint them while building, it will be a very difficult job to paint them later. And you do want to paint them with hard enamel paint, for a hard reflective surface. I have seen some for sale online that are not pained - not a good idea! Every time you glue, make sure it is square:

Here are some pictures of the ones I built:

Diffusion: Skyline

SKYLINE DIFFUSERS ARE ANOTHER OPTION, where the DIY build process is achievable. It is a backboard on which you glue square pine wood of different lengths, sequentially placed. The materials are simple and obtainable from your local home depot. You'll just want to get a small blade drop saw for cutting the lengths and angles:

REF: *Image source unknown*

Here is the BBC technical document on Skyline Diffusers, by R. Walker, see <u>downloads.bbc.co.uk/rd/reports/1990-15.pdf</u>. Here is a good YouTube DIY instruction video, by FocusOnline, see <u>youtu.be/qFeM2uWuMZI</u>. The two downsides are that, (a) they will be very heavy, and, (b) you will need to be about twice the distance away from them as from the QRD diffusers, so these are more suitable for medium to large sized rooms.

Diffusion: Fractal

FRACTAL DIFFUSERS ARE ANOTHER OPTION, where again the DIY process is achievable. These are also only suitable for a large sized room. Here is the Arden diffuser, with access to the blueprints and DIY build information, by Tim Perry, see http://arqen.com/sound-diffusers/faq/:

REF: Image source from site referenced above

THE RESULTS OF DIFFERENT ACOUSTIC TREATMENT

ABOVE, I showed a waterfall image of my recording space. Here is another graph of the same space, up to 300Hz, showing the effect of the different acoustic treatments:

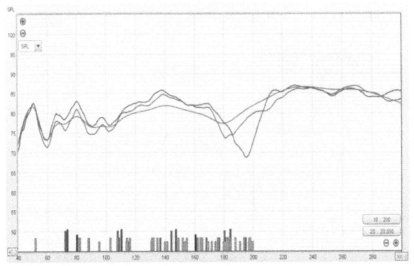

First, <u>the line that dips down the most</u>, is achieved with **4-5" broadband panels and ceiling Superchunks** up in the ceiling-wall join on each side of the room. Then <u>the line that dips down less,</u> is achieved by adding **diaphragmatic bass traps,** side, back and front. Then, the <u>flatter line</u> depicts the advantage of diffusers on the back wall and one side wall.

The flatter line, the final result, is as good as I could wish for such a small space.

Yet, let us consider what could be done to perfect it. The dip at about 60 Hz could be exacerbated by the plasterboard walls, or the design of my membrane bass traps. 12.5 mm plasterboard on 75 mm studded wall cavity will resonate at around 60 Hz, which could be pulling too much 60 Hz energy out of the room. Two sheets of plasterboard will resonate at around 50 Hz. So, if I add that it should bring down the 50 Hz peak, and perhaps it will also bring up the 60Hz dip. I could lean another layer of plasterboard against the walls either side and see if I am right. If so, I might pull off the plasterboard and insert acoustic wall batts, then add back two layers of gypsum, gluing it together with acoustic level polyurethane adhesive sealant (the fire retardant grade). I am not going to do this, though. I am happy enough with the present result, for the time being.

 PLACEMENT OF ACOUSTIC TREATMENT

FOR PLACEMENT, you can see the suggested way acoustic treatment is positioned for greater effectiveness – the highest amount of acoustic treatment being for recording spaces. For a listening space, the first or middle amount would be fine because, applying subtractive equalization, surround sound, multiple sub-woofers, and opening doors and windows if you can, contributes to the efficiency of acoustic treatment. As mentioned above, as the room gets smaller, keep the same amount of treatment to remain effective.

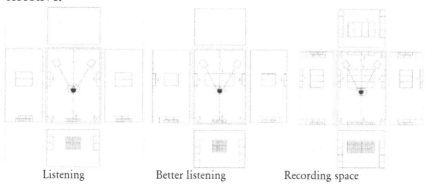

Listening   Better listening   Recording space

Images from Master Handbook of Acoustics (6th ed) p744-750, by DF. Everest and K. Pohlmann

SUMMARY

AFTER APPLYING ACOUSTIC ROOM TREATMENT, direct reflections, room modes, and reverberation/echo will have been tamed so as to enable enjoyable visceral audio and uncompromising recordings. You will notice this improvement by clapping, playing music, or simply talking in the room; it will be clearer, crispier, and tonally more pleasing, and more relaxing. You will hear the audio clearly, and feel it!

The degree that this is achieved depends on how far you are able to go with acoustic treatment. For example, as discussed above, in a professional studio setup where there the room size is also preferable, an acoustician would apply a second application of custom designed acoustic treatment after measuring the room, such as by building targeted membrane bass traps. This takes more space and time, and is generally not pursued for home recording and home listening spaces, but, we have seen how this is done, should you wish to do it.

I have also shown how with decor considerations, you can apply quite a lot of acoustic treatment in a lounge listening room. In the chapter on 'lounge listening' we also explore speaker design and placement, and subwoofers.

CHAPTER 4
RECORDING SPACES

"I also have a recording studio that I use to produce bands, by Steve Brown.

IF YOU WANT TO RECORD ANYTHING, this chapter will help you choose the right kind recording space strategy for your house or apartment. Also, if you understand the recording side of things then it helps to understand the listening side of things; two sides of a system.

While it is not a creative enterprise, in order to create, we need a good environment to capture that creativity, so you can share it with others who can truly connect with what you produce; it is a tangible, necessary element of recording. So then, we need to know the kind of spaces that are good for recording. We began, as a first step, to think about acoustic treatment. Now we will consider some visual and audio components for recording. Not just the technology, but the idea that we should look our best, and actually look at the people we are talking with.

CHOOSING A RECORDING SPACE

BECAUSE OF OUT-OF-HOUSE SOUNDS, we want as quiet a room as possible in the house, with the following considerations:

- An internal room with **one wall to the outside world will probably be nice.** If you have a basement with concrete or brick walls, it might seem like the best option because of the quietness, and it would be if you are in a flight path or near a busy road, but then you will have to do more extensive work on bass trap acoustics to control the bass sound bouncing off the concrete walls. It depends on your needs, but it may be better if you have normal partition walls, where some low bass will rumble through and so requires less bass trapping – hopefully external sounds won't intrude during recording times.
- **A bigger rectangular room** is acoustically better than a square and/or a small room. This is because the internal sound bouncing off the walls is worse in a square configuration, and, a bigger room provides space for the energy dispersion.
- Even though a square and/or small room is bad acoustically, **nothing is end of the world.** In fact my lounge is almost square - I will show you what you can do to improve it.

 ## A SMALL RECORDING SETUP

I ONCE BUILT A DEDICATED RECORDING STUDIO (shown below), partly because I wanted to be isolated from outside noise, because we were not far from a flight path, but, while it was an excellent place to record in, when we moved house I had to leave it all behind. In our new house, I have been intent on making mobile acoustic treatment, so if we move again we can take it all with us.

For my present recording space, I am limited. I use part of a long rectangular room, which feeds into a bigger L-shaped room. That isn't perfect, but it is often the case that you will need to make some compromises. For example, an ideal room will have a high ceiling, yet that is more than likely not what you have. At the moment I am using this room for on-line teaching, and for off-line recording, so I have also paid attention to the lighting and video setup (which I cover below). In my case, I have a standing setup, because having been a teacher for a long time, I always found I have more energy if I stay on my feet.

LET'S SEGUE TO THE VISUAL ELEMENTS:

Lighting? I have a five light setup: One light is shining from back-above onto my shoulders for highlighting, one light from the side-back for back light, and one from side-front, highlighting walls, to fill the space with light and get the lighting right on my face. Then, I have a four-bulb-tent-light behind my screen shining up, to reflect disbursed light from above and add the right amount of light to the front of my face. There is no direct hard light on my face; it's all diffused. Therefore, I do not have any **light reflections in my glasses** - apart from what might reflect from a screen I am looking at (there is a solution to that too).

For online sessions, you will want your viewers to see your eyes. I wear glasses to read the screen, but even with non-reflective glasses you cannot normally avoid a direct light reflection from an open document on your screen. This is because the on-screen open document has a while background by default. You can overcome this by:
a) using an external screen, so it's not full of your usual icons.
b) make your background screen color black or dark.
c) make your open document page background off-white, such as with these settings in Libre Office Writer:

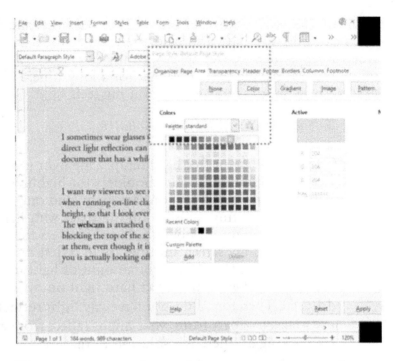

There is one more issue with an external screen, that of the audio reflections it can potentially cause. To address this issue I have the external screen up higher than my mouth, so that, a) audio reflection are directed upward as as opposed to directly into the microphone, and b) some direct vocals are absorbed by a bass trap. Having your camera so you visually look slightly up to it is also a more flattering angle visually. I then keep the actual computer lower down, with the screen tilted down, so my vocals are not reflecting off that:

For other purposes, you may be able to use a Tablet rather than an extended computer screen, which reduces the potential screen reflections further.

My **webcam is** attached to a **ceiling mount** and **positioned over the top of the screen** - yes, partially blocking the top of the screen. This way, when I read or talk to people on-line it looks like I am looking at them, even though it is just a little annoying for me:

The impression that you are looking at the person you are talking to is super important for effective communication. If someone who talks to you is actually looking off to the side – how does that feel? The distance you stand from the camera should be a normal social distance of about three feet, where the viewer can also see hand gestures.

Webcams do have one big issue, if you are going to use them for a solo web session. They have wide angle lenses, suitable for web conferencing, or if you are demonstrating something in a room, but not for a talking head approach. The wide angle lenses of webcams make your face look fatter than it really is. I don't think many people want to be seen like that. Comparatively, a portrait lens gives the face and body, real, harmonious – even flattering – proportions. There are a couple of ways to set up a portrait approach for streaming.

One way is to use an additional lens, a portrait 58mm lens, over a Logitech C270 webcam. It makes the world of difference, see the Moment Tele Lens, www.shopmoment.com/products/tele-58-mm-lens. I use this mount, www.shopmoment.com/products/m-series-lens-mount, but snapped off the phone clip part, then glued it over my webcam lens. It works well. The image quality can be bettered, but I prefer a flattering portrait image:

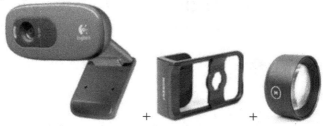

Without the benefit of a portrait lens, the best webcam I can presently see on the market is the Razer Kiyo Pro Ultra, see https://www.razer.com/ap-en/streaming-cameras/razer-kiyo-pro-ultra. See this review by EposVox, https://youtu.be/vCwAEirOoqo.

Another way, and for a much better portrait picture than a webcam can provide, is to use a mirrorless camera, and attach a 50mm lens or a zoom lens with a 50mm range, see www.thephoblographer.com/2022/02/03/is-35mm-or-50mm-better-for-portraits-we-take-a-look/.

I do suggest a zoom or variable lens for flexibility, to handle small movements on autofocus, and if you want to use your camera for other purposes. In the Sony range, look into the **Sony A7 III camera,** or Sony A7SII for a cheaper option (which has a little slower auto focus). You should be able to get a good second hand camera and lens, or on sale.

If using the HDMI connection (recommended), also get a **USB to HDMI capture card.**

For the add-on lens, here are some options as recommended by Abby Ferguson at popphoto.com, see www.popphoto.com/reviews/best-sony-lens-for-portraits/. It should have a short minimum focal length, about a foot (0.3 to 0.4m) close enough that you can mount the camera from a ceiling mount, just in front of an extended screen (for a direct gaze), and good auto-focussing. Keeping those options in mind, the best value for money seems to be the **Tamron 28-75mm f/2.8 Di III RXD Lens (Sony E),** with a close option being the **Sony** Vario-Tessar T* FE 24-70mm f/4 Lens.

Suggested: a ceiling mount setup, Sony A7 III (or A7SII as a cheaper model), Tamron 28-75mm f/2.8 Lens, USB to HDMI capture card.

Here is a video for setting up one of these Sony cameras, by Think Media, see youtu.be/fj3qxLRH-Po? t=152. As mentioned in that video, you can plug in a USB microphone, like the Rode VideoMicro, see rode.com/en/microphones/on-camera or similar.

However, in an acoustically treated room you will get far better audio through a separate digital interface and a microphone that needs phantom power (powered from the digital interface) – suggestions are discussed in detail in the chapter on 'Microphones, Digital Interfaces, and Plugins'. I only recommend using a Rode VideoMicro, or similar, for shooting video outside where plugging into power and a digital interface is not feasible.

For a ceiling or wall mount, as opposed to a desk mount, these bits would do the job:

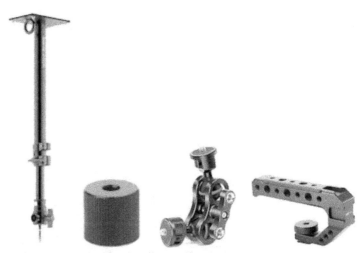

- Selens Photography Studio Wall Mount.
- 1/4" Female To 1/4" Female Universal Convert Screw Adapter.
- UTEBIT 360 Degree Rotation Magic Articulating Arm Double Ballhead Arm 1/4".
- UURig R005 Camera Hot Shoe Top Handle Grip.

You can also connect these cameras using USB rather than HDMI, in which case you do not need a USB to HDMI adapter. But, you will in that case need the software (such as provided by Sony). I prefer not to recommend that, to avoid the complexities and potential issues of requiring the extra software.

The basic camera settings to get going for the Sony A7 III are Top dial=M, Quality image=JPEG, Record Settings=60p 50M (change to 25M if streaming is slow), Steady shot=On, Menu>HDMI Settings>HDMI Info Display=Off>, Face/Eye Prty in AF=On, Focus Mode=Continuous AF, File Format=XAVC S HD. The following is for a different Sony camera, but this is good advice to understand the settings involved, by The Three Techs, see youtu.be/C1Nvp_46ZO0.

NOW, LET'S SEGUE BACK TO ACOUSTICS:

I have said I use a bar stool and standing position for recording and running Internet teaching sessions. This picture, below, shows the off-screen microphone setup, which I do for Internet sessions. I bring the microphone setup closer for non-video recording. Above the recording position, I have a thick 6" absorber, and where the ceiling and wall meet there are some super bass traps (building instruction for all these acoustic treatments are in the chapter, 'Room Acoustic Applications').

Here are some more pictures of the surrounding acoustic treatment:

The left picture is behind the recording position, some bass traps with diffusers stacked on top. The middle picture is facing forward at the recording position: directly in front there is a bass trap at my standing mouth level, to the right there is a bass trap and above it there is a broadband panel. The picture on the right is to the other side of the recording position, and includes bass traps and QRD diffusers.

You can see there is a lot of acoustic treatment jammed in there. This is what happens when the space is small. If it was a bigger space I would have the same amount of treatment, spread out more.

The sound made in this space is clear and crisp.

There is some more discussion of this setup in the chapter, 'Room Acoustic Applications'.

 A PERMANENT HOME STUDIO WITH AIR FLOW

BELOW IS THE PERMANENT HOME STUDIO I built in our house in Brisbane, Australia, where I could build into our large rumpus room. Because I wanted to spend hours each day recording, and because there was an international flight path close enough to cause problems, I ended up building this as an internal studio space, taking up about a third of our large rumpus room. It was quite an undertaking.

If you are in a situation where you want to build a studio, my advice is to contract a professional adviser, and do as much DIY as you want to, need to, and are able to. I recommend John Brandt, see jhbrandt.net/. Apart from there being various free DIY resources on his site, he offers to work with you remotely, through the process of your build. However, I did it on my own, taking about one year of research and building, on top of my daytime job, which is why I make that recommendation.

It was a room-within-a-room design. This means, it has double layer, not touching, internal walls. From outside in, to make the walls: there are sandwiched layers of Gib rock wall panels on the outside (glued together with a flexible polyurethane sealant), these, on a wall frame filled with wall insulation, then a non-touching air-gap, then a second layer of wall framing, also filled with insulation. Instead of Gib on the inside wall, there is an air gap, then a facing of mass loaded vinyl (MLV) stapled to the framing. As a finish, I put wooden slat spacing over the framing, then hessian fabric covering – thus making a substantial part of the walls a bass trap. The room also has a double-glass-pane window, with an air space between the class panes, the inner glass angled downwards to mitigate reflections. It has a double doors with an air space between the two doors, attention paid to a good seal on the doors. The whole wall partitioning is separated from the concrete floor by a rubber footing. The ceiling has insulation is built into it, covered with hessian fabric. With the doors closed no sound can be heard from outside, except thumps on the floor above.

One big issue with a studio like this is how you are going to breath. You either have regular breaks or you need a way to get air into the room in a soundproofed way. I pumped air in from the larger rumpus room, which was air conditioned. It worked well. So, for air in, a ducting fan was connected to an insulated plenum sound box, connected through to the studio's rear wall via insulated ducting, at which point the air path was fed into another large plenum (four pictures below), and finally into the studio. Some of the pieces looked like this, doubled up; one system for air in, the other for air out:

Below, with the rumpus room cupboard open (left of picture), and with the doors shut (right of picture) we see the arrangement.

The bottom grating is the air inlet; the top-right grating is the air outlet:

Below, I show the building design of the studio side air ducting. You can see that the internal ducting is fed into the two larger plenum sound boxes providing the studio-side sound separation path into and out of the studio. I painted on the outside to show the air pathway:

Below, is the inside studio back wall, complete. The result was, less than 1 dB of sound at the internal air gratings in the studio, yet fresh circulated air! You can see the lower air intake grating near the floor, and the air outtake near the top. The brown hessian fabric rectangles are covering deep base traps. The wall carpet has insulation behind it, the hanging plastic leaves is a cheap way to diffuse rear wall reflections:

Then we moved to New Zealand, and in doing so I had to leave all that work behind. Bugger! It is a good idea to build a studio if your location is permanent and you need to be isolated from outside sounds. Otherwise, temporary mobile treatment is the way to go, as discussed in the next chapter.

PORTABLE BOOTHS

A PORTABLE VOCAL BOOTH is only really an option for a double garage space, where there are no other options in your house. The best resource for building your own portable booth, that I have found, is from John Brandt, see jhbrandt.net/resources/. The at present $100 fee is a steal. I have above, also suggested using John Brandt if you do want to access a consultant, either to help building this portable booth or a larger studio room. John Brand has provided a lot of free resources for the DIY community, so this is my way of saying thank you. He pays me nothing for this recommendation.

Do not try and build a vocal booth into a bedroom, instead build acoustic treatment that can be moved as needed. Much is explained on this subject in the chapter on 'Room Acoustics Applications'.

You can buy, portable booths of the type made up with framing and acoustic blankets, but for a room, it's just the wrong way to go. Acoustic blankets will never combat bass room modes, the worst culprits for recording.

SUMMARY

WE HAVE CONSIDERED what kind of spaces are good for capturing your creativity in order to share it, how to build a recording studio or portable booth, but I concluded that 'modular' acoustic treatment is a flexible strategy in case you move premises. I also segued to some visual and audio components for on-line recording, which so many of us these days are involved in, bearing in mind our interest in looking our best and looking at the people we are talking with from a comfortable social distance..

CHAPTER 5
ISOLATING A ROOM FROM OUTSIDE SOUND

The condition every art requires is not so much freedom from restriction, as freedom from adulteration and from the intrusion of foreign matter", by Willa Cather.

THIS CHAPTER DELVES INTO HOW TO ISOLATE YOUR RECORDING ENVIRONMENT. Apart from choosing a quiet place in your house to record, you may need to isolate yourself from outside noises, and/or isolate others from the noise you are making. It solves the annoying problem of the intrusion of outside noise into your creative process. Perhaps you are living on or near a busy flight path, a busy train track, a motorway, or main road, or, you are making heavy metal or rock music. Before thinking of recording booths, look at what you can do to isolate a room, because the larger the room size is, the better the recording is, and the better the listening environment is.

DOORS

THE WEAKEST LINKS ARE DOORS AND WINDOWS. Doors will be the weakest link of a room to the rest of the house; windows to the outside world. Internal doors are typically very thin, many with a cardboard-like internal honey comb separation between thin door panels. One easy way to block external noise from coming under the door is to attach a draft stopper, some of which can slide onto the bottom of the door. A more permanent solution is to attach an external door sweep to the foot of the door, one that is more solid, made of rubber. Usually, these are for external house doors, but you can put them on internal doors for acoustic isolation purposes. An additional solution to really solve the door problem is to buy a solid door and swap it over. These can be hinged on existing hinges, so even if you are renting, you could swap back the original door if you move.

Ref: *https://www.soundproofingstore.co.uk/how-to-soundproof-a-door*

WINDOWS

WINDOWS, IF SINGLE PANE, LET IN SOUND MORE THAN OUTER WALLS. The more expensive way is to replace single pane window/s with double or triple insulated glass. While an excellent option, that is expensive, and you may not be in the position to do it.

Alternatively, and depending on your window frame design, you can get acrylic sheet/perspex, or glass, cut to the size of your window glass, and slot it in with a gap, ideally a 12mm gap. You can cut acrylic sheet with a jig saw or skill saw, slowly with a bit of hand-soap added where the blade is cutting, and using a metal cutting blade. You can usually get acrylic sheet at a home depot, and glass from a glazier. Here, at Ephe - Residential Energy Efficiency, we see how this can be done, see youtu.be/hgb-wL_VsGM. To prevent condensation, I would add some small air holes drilled through the window ledge shelf, or top, bottom, or side, and leave some silica gel beads in the gap.

To seal it in a way that can be removed later, you could use temporary caulk, that can be pealed off, something like this, www.dap.com/products-projects/product-categories/caulks-sealants/specialty/seal-n-peel/.

Acrylic or glass allows you to see outside, which, depending on your view, may be important to you. If the view can be compromised, an alternative method is to make a window plug, the benefit being that it becomes part of your acoustic room treatment. A window plug is basically a broadband panel made to snug fit your window crevice. If it tends to fall out, it is better to add some material to thicken the snug fit, because any air gap is a sound gap. While you are at it, have a look at your window frames, making sure there are no air gaps. If there are gaps or cracks, fill them.

EXTERNAL WALL

IF THE EXTERNAL WALL IS NOW THE WEAKEST SOUND LINK to the outside world, you can check if it has insulation built into it. If not the best thing to do would be to take off the existing plasterboard, fill the gaps with wall insulation, then cover with new plasterboard, better still, two layers of plasterboard glued together with acoustic sealant. If you are renting, you could try some moving blankets, hung by hooks, or screwed on. You can get them from home depots or moving companies. It can act like a super thick curtain - the weakest link, however, is any air gap, so pay attention to the edges.

CEILING, IF A HOUSE WITH A ROOF

IF YOU CAN GET UP IN THE RAFTERS, add household insulation, cut to size for the gaps between rafters.

FLOOR

IF NO CARPET, ADD A CARPET TO MITIGATE FLOOR REFLECTION, particularly between you and any desk, preferably with a layer of wool underlay.

STILL ISSUE WITH INTERNAL WALLS?

IF THERE IS STILL AN ISSUE, even an issue of your own noise seeping to the rest of the house, you may need to get serious with your walls. In this case you need to be the owner. The easiest thing to do would be to add a second, even third layer of gypsum, gluing the layers with acoustic level acrylic sealant, paying attention to the edges. If you are going to that trouble you might as well cut out some holes between the studs and put in some acoustic panels if not already installed.

 AIR

IF YOU HAVE DONE A LOT OF THE ABOVE YOU WILL SUFFER FROM A LACK OF AIR if in a closed room for long periods. First off, you can take regular breaks with the door open. The next option, if you have sealed up the room, is to pump air in and out. This is not so easy because you need to do so in a soundproof manner. If you can get away with it, the first possibility is a split system air conditioner, one with variable speed so you can run it slowly and quietly. This could work if you are recording with a shotgun or dynamic microphone, but not otherwise.

I did make an air system that only had 1 decibel from the sound of air at the intake (see the chapter on 'Recording Spaces and Room Acoustics'). I can tell you it is a lot of work, but you could achieve it by using a wardrobe space and an inlet and outlet into a hallway. You would need to build 4 plenum boxes, two of which have air duct motors with variable speed, and two of which lead the air to travel around corners from one end of the box to the other. These sound boxes need internal high density insulation against the inside surfaces.

SUMMARY

TO THE DEGREE YOU HAVE ANNOYING EXTERNAL SOUNDS intruding into your recording space, these applications will need attention; likewise, if you are bashing out music and want to be considerate to other people in your house or apartment and/or your neighbors.

CHAPTER 6
MICROPHONES, DIGITAL INTERFACES, AND PLUGINS

"The tone is the message", by Kevin T. McCarney.

TO RECORD SOUND, YOU NEED a microphone suited to purpose, a digital interface to be a lot better than a computer's sound card, and headphones and/or monitors (reference speakers) so you can monitor the details of what you are recording, and edit and/or mix, and master with confidence. By doing that you are recording tonally clear material, digitizing it without adding noise.

At this point we can begin to consider the issue of digital versus analog recording. The whole discussion about analog sound versus digital, to me, revolves around the emotional connection derived from the tactileness of analog mediums. I imagine walking on the beach, yet with the sound of feet on concrete instead of sand, or the gently crashing waves, yet slapping on concrete instead of flopping and dragging on sand. There is a subtle difference in tactileness, which is emotionally felt. The argument for analog is then a lot about the tactile visceral audio connection of the physical elements involved in recording (and playback) that creates a slightly different analog tone, which likewise creates a sense of emotional connection. While digital technologies do have connection at a physical molecular level (usually copper), other mediums like ribbons in ribbon microphones, like tubes, like tape, like resistors (which have powdered carbon) create a natural sounding tonality. I keep this in minds as I explain microphones, digital interfaces and the plugins that are usually involved in the recording process.

MICROPHONE SETTINGS

BEFORE TALKING ABOUT MICROPHONES, I will introduce the key settings for recording good sound, listed as 1-6 below. To apply these settings, You can use your DAW and plugins, or go old school with an analog Channel Strip device (often involving tube technology) like the Avalon 737sp, see avalondesign.com/vt737sp.html, which can be plugged in between your microphone and digital interface. The settings on this device represent key settings you need one way or another:

Ref: Image from site referenced above

The settings are:
1. 'Preamp gain'. Your Digital Interface will have preamp gain. You set it to at the level where there is no clipping (no flat heads on the recorded wave peaks).
2. 'High gain', boosts volume, with more dramatic tone control.
3. 'High pass', is a setting to roll off low bass sound. This is useful if you are recording dialog or song ('high pass' means letting high frequencies through/blocking low frequencies that are not needed). There is no point in recording any unnecessary low end rumble or noise. Sometimes on the actual microphone, there will

be a 'high pass' switch (Sometimes they call that same switch a 'low pass filter', so check the manual).
4. 'Compressor' settings, lowers high volume peaks, thus increasing quiet volume frequencies, making sound clearer to the ear when the listener has their own competing background noises, like car noises. Some music,(compression does reduce the natural dynamic range between the highs and lows). targeted for playback on headphones and or a home hi-fi, need not have as much compression. Compression does reduce the *natural* dynamic range between the highs and lows.
5. '+48v', is needed for condenser microphones (sending power from the computer to the microphone circuitry).
6 'EQ' controls, compensate for room mode and echo issues, and can also shape certain frequencies, which certain instruments benefit from.

HARDWARE UNITS, like the Avalon 737sp (which uses tube pre-amplification and optical compression), are less common these days, because Digital Signal Processing (DSP) settings in quality digital interfaces (DIs) and software plugin alternatives are available for a much cheaper price, often simulating the analog tonal character of older classical analog equipment. By the way, an optical compressor uses a light element and optical cell to alter the dynamics of an audio signal, creating an elusive euphonic warmth and musicality.

 GETTING ADVICE ON A MICROPHONE

IF YOU WERE TO TAKE PART IN A RECORDING SESSION AT A PROFESSIONAL STUDIO they would test out a number of microphones to find one that suited your voice and music genre. So, you really need a professional who has a wide range of knowledge of what microphone would be suitable. That may not be possible for you. What we are trying to do is match the sound signature, including tone, of your voice, with the sound signature of the microphone, in a complimentary way.

One way to find the right microphone, is to ask for advice online. You could get a commonly known budget friendly microphone, make a recording, then seek feedback on Gearspace>Electronic Music Instruments and Electronic Music Production, see gearspace.com/board/electronic-music-instruments-and-electronic-music-production/ or other forums. For this purpose, upload the recording to the forum, without applying any effects, and ask for microphone suggestions. Include, your purpose, your budget, the name of any well known singer or narrator that you would like to sound somewhat like, and the microphone and equipment used to make the sample. Hopefully, as I found, a number of professionals will offer you good advice. As already stated, if you have access to an in-person audio engineer, that of course would be the best idea. If you are thinking of using an audio engineer to mix and master your work, then why not reach out to them for this kind of advice.

For creating a sample, a well known relatively cheap microphone, could be one of the Blue Yeti USB microphones, see https://www.logitechg.com/en-us/products/streaming-gear/yeti-premium-usb-microphone.988-000103.html.

It is a USB condenser microphone, and so is quite sensitive to room echo, so if you do not yet have acoustic treatment, record while facing into a closet full of clothes. It is connected by USB, so you can plug it straight into your computer (Windows or Mac).

Another way to find the right microphone is to use the Townsend Labs Sphere L22 modeling system (discussed below).

POP FILTERS ARE NEEDED FOR VOCALS

IF YOU ARE SPEAKING OR SINGING INTO A MICROPHONE DIRECTLY, you need a pop filter (unless built in), to prevent distractive plosives from being over-stated in your recording. Plosives are abrupt air expulsion, as in p, t, k, b, d, g in English words. Highly recommended, is the Stedman pop filter, see stedmanusa.com/pop-filters. Here is a comparative review, by Mike DelGaudio, of Booth Junkie, see youtu.be/amWbTkjfhDk:

Mesh style pop filters take something away from the analog tone of your sound.

 ## DYNAMIC MICROPHONES

DYNAMIC MICROPHONES CUT OFF THE LOW AND UPPER FREQUENCIES that are outside of the range of the human voice. Also, they use a cardioid polar pickup pattern, where the sphere of focus for the microphone pickup is wide (front and to the sides). This pattern allows for some creative movement as you record, while minimizing any other noise. Remember we are after your direct recording and just a little bit of room ambiance:

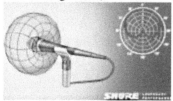

Think, *"on stage, radio broadcast, interviews, and Youtube channels"*.

They will do a good job of cutting out outside-of-house, and outside-of-room noise, which may otherwise intrude into your recording. This is why they are used for live stage work and interviews.

Technically, dynamic mics are passive, meaning the audio signal is not increased until it gets to a digital interface, specifically, the preamp inside your DI, or in the case of USB dynamic mics, the preamp inside your computer.

One well known dynamic mic to consider is the Shure MV7, about $450, see www.shure.com/en-US/products/microphones/mv7?variant=MV7-K:

The Sure MV7 is based on one of the industry standard interview mics, the SM-7B, see this review by recording hacks, see recordinghacks.com/microphones/Shure/SM7B.

The MV7 comes with an App for adjustments, and, as well as the USB connection, it has an XLR connection, so initially you can plug it straight into your computer via USB, then later, for better recording quality, into a digital interface which will have XLR input connectors. The MV7 and the SM7B do not need a pop filter, because it is built in.

Although preferred for vocals, it can still be used for different purposes: vocals (15 cm from mouth); Guitar & Bass Amplifiers (60 to 90 cm back from speaker), as per the guide. Oh yes, Michael Jackson used one!

See also the RODE NT1 5th Generation Condenser microphone below for a similar USB connectivity option.

LIKEWISE, and another mic that you talk into the end of, and one that is often seen on YouTube interviews, is the Electra-Voice RE-20, for about $1,200. Here is a review by Recording hacks, see recordinghacks.com/microphones/Electro-Voice/RE20. Again, it's used for vocals, think, *"close intimate sound"*, for music, think, *"rock n roll"*.

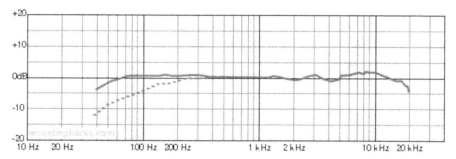

Again, this one, like the MV7 also has a built in pop filter. The frequency response chart above, shows how evenly true to life, it records, all things room acoustics being equal. The dotted line shows how the bass can be rolled off if you engage the high pass filter switch (designed to take your voice out of a listeners sub-woofer), and to remove the sound not covered by the human voice, because there is no need to record any possible low end rumble or background noise. The RE-20 is sometimes suggested as a better choice than the SM-7B because it is more versatile for both male and female voices – it is better at capturing the higher frequencies of the female voice. The RE-20 only connects to an audio interface, which then connects to your computer. Here is the data sheet:

products.electrovoice.com/binary/RE20_Engineering_Data_Sheet.PDF. Yes, used a lot for podcasts. Oh yes, Stevie Wonder uses one!

FOR ON STAGE WORK, perhaps the world's most popular performance microphone is the **Shure SM58.** It has a Cardioid polar pattern is the Microphone. Think any artist at any live venue, in the last 20 years. It is quite possible they used a SM58 for their live performances. It's cheap, too. Oh yes, Paul McCartney uses one!

Or, if you want to go wireless, try the **Telefunken M80.**

Again, the Telefunken M80 is a dynamic microphone, but this time with a sound more like a condenser sound, "*crisp, accurate, clear, and bright*". There are many color options, too.

IF YOU GET A DYNAMIC MIC, because they are passive, you might not get enough '*clean*' (no hiss) volume from it, although, this is less likely if you have a quality digital interface. On a quality digital interface you should be able to turn the gain/volume knob right up without any hiss. If you are struggling to get the volume you want, you can add a preamp to boost the volume between the mic and the audio interface, one common one being the Cloudlifter CL-1, about $250, see www.cloudmicrophones.com/cloudlifter-cl-1:

I have a Grace M101, which is very 'clean', meaning there is no introduction of its own tone. I use it with a ribbon microphone, utilizing its ribbon mic mode:

As a reminder, WITH DYNAMIC MICS you can get away with minimal room acoustics.

PROXIMITY EFFECT is when you speak close up to the microphone, sometimes for an authoritative or whispered sound. The Sure MV7 and RE-20, and all quality microphones, are designed to work well when you want to get that up and close proximity effect, whereas cheaper microphones will most likely have a boomy proximity effect.

YOU WILL PROBABLY NEED A SHOCK MOUNT. With all microphones, apart from dynamic mics designed for stage use, you need a **shock mount** to eliminate vibrations. For the RE20 get this one, for about $100, see www.bhphotovideo.com/c/product/76703-REG/Electro Voice 3091965 309A Suspension Microphone Sh ock.html.

The Sure MV7 comes with a shock mount in the box.

You might also want a **boom arm**, like this one:

www.amazon.co.uk/YOUSHARES-SM7B-Boom-Arm-Filter/dp/B092HPD397

 ## CONDENSER MICROPHONES

CONDENSER MICROPHONES ARE MORE SENSITIVE than dynamic mics. Even though they usually have the same cardioid polar pattern as the dynamic mic, they pick up more ambient sound – which is higher and lower than the human voice. It is generally better to include the tonal atmosphere of a room, which encapsulate the musical resonances. In particular, this will provide a better recording of singing, and musical instruments, the higher quality mics giving you better room atmosphere, and being more likely to work with multiple voices.

Because of the sensitivity of condenser mics, you will need room acoustic treatment. And you will need a digital interface, to feed it with 48V phantom-power from your DI.

The Rode NT1, for about $200, is a very popular mid-priced singer's choice condenser microphone. It is not expensive but reliably good, see rode.com/en/microphones/studio-condenser/nt1. Think John Lennon, Rod Stewart, and Billy Idol! If you do want the type of USB connectivity option as the Sure MV7 above, there is an XLR and USB version of the NT1, see the Rode NT1 5th Generation, see rode.com/en/microphones/studio-condenser/nt1-5th-generation:

 SHOTGUN (CONDENSER) MICS

BUT, BEFORE THINKING THAT *"SERIOUS"* ACOUSTIC TREATMENT IS INEVITABLE for a condenser microphone, there is one good enough work-around, especially if you are recording podcasts, videos, voiceover or narrator work. **Shotgun (condenser) mics** were conceived for movie sets and outdoor recording - focusing a narrow pick-up area so they can be distanced out of the video shot, and from external noise. Therefore, we get a quality recording but one that is focused so as to cut out external noise, and out of view of a webcam. It targets the vocal area really nicely from a foot or more distance. So, we can get away with minimal acoustic treatment and get the extra quality of the condenser technology, *"accurate studio recording: crisp, accurate, clear, and bright"*. Quite a good solution. Actually, having extensive acoustic treatment will still sound better with a shotgun mic, because room echo still intrudes into all space, but it can be good enough with minimal acoustic treatment.

FIRST WE WILL LOOK AT A VERY VERY REPUTABLE MODEL, then a very similar Chinese made model. The **Sennheiser Shotgun 416** is a very well known microphone for voice over and narration work. Here is a review by Podcastage, see youtu.be/hI3YW7DjOk8, and by Rebel Tech, see youtu.be/yBqu3qKP2rk, and by Brodie Brazil, see youtu.be/swdJmrO7WiU.

Here is the Sennheiser Shotgun MKH 416-P48U3, at about $1,500, see en-us.sennheiser.com/short-shotgun-tube-microphone-camera-films-mkh-416-p48u3:

IT IS A BIT CRUEL TO CALL IT A KNOCK-OFF VERSION, when referring to the Synco Mic-D2, at about $350, see www.syncoaudio.com/products/sy-d2-bk – but it is modeled off the MKH 416. See this comparison review by Booth Junkie, see www.youtube.com/watch?v=NmJeMzl0JZI, which led me to buy one.

THE POLAR PATTERN of a shotgun microphone is *super-cardioid*, which, enhanced by the tube design, rejects noise from the sides. One negative is that if you are close to it you cannot move around too much without causing volume fluctuations:

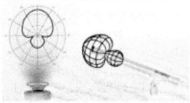

If you do want to get up close to it, for an authoritative / proximity effect, you could get this pop filter - the Hook Studios Octo-842S, see thehookstudios.com/filters.html.

If you are shooting video outside, you might need a windshield, see mymic.rycote.com/products/windshield-solutions/classic-softie-camera-kit/18cm-classic-softie-camera-kit/.

 ## RIBBON MICROPHONES

RIBBON MICROPHONES USED TO BE 'THE' MICROPHONE for music and vocal recording, think, 'Elvis Presley', before the invention of the Condenser mic. Think, *'silky and natural'*; think analog.

As analogue recording devices, they are still one of the best for picking up tonal nuance. For example, the tonal character of different guitars may not be noticeable to a dynamic or condenser mic, but it will be picked up on a Ribbon mic (or Tube mic). This also applies to the nuanced tonal nature of the spoken or sung voice, think, *'emotional tone variation'*; think analog.

They are often passive, meaning they need an external preamp before the digital interface. Usually ribbon preamps have an impedance switch (balancing the voltage and current), for freer movement of the ribbon; thus a full sound.

They are very sensitive, picking up any external sound, and can be damaged if dropped, so are not as commonly used in home recording, often voted out by a preference for the accuracy of condenser mics, and because of the extra cost of a specialized preamp. A setup would look like this:

PASSIVE   PRE-AMP   DIGITAL INTERFACE
RIBBON MIC

I have a Sure KSM313/NE, which costs about $1,900, see

shure.com/en-MEA/products/microphones/ksm313?
variant=KSM313%2FNE), and a Grace M101, about $850, to
preamplify the signal, see gracedesign.com/products/microphone-
preamplifiers/m101:

The Shure KSM313 is designed to have different sonic
signatures from the front and rear. The front of this microphone
is used for electric guitar amplifiers, brass and woodwind
instruments, kick Drum; the Rear side for vocals, acoustic string
instruments, drum overheads, and percussion.

The ribbon of the Shure KSM313 is made of a hybrid metal, so
it is not susceptible to breaking if the mic is dropped.

THE GRACE M101 is a clean preamp, meaning it does not color the
sound. It is of such good quality I can crank up the gain/volume
to 100% without getting any hiss.
    Other quality ribbon preamps include the AEA TRP2, about
$1,000, see aearibbonmics.com/products/trp2, the Cloudlifter CL-
1, about $300, see cloudmicrophones.com/cloudlifter-cl-1, being
more of a budget option. For musical instruments, the Cloudlifter
CL-Z, offers settings to adjust the impedance for sculpting the
tone, as explained in this video by Cloud Microphones, see
youtu.be/gqxNCVJI8DI.

HERE IS A DEMO OF SOME OTHER RIBBON MICS used for vocals, by Andertons Synths, Keys and Tech, see youtu.be/Q6FnlI4jDAU.

WELL, THERE ARE BUDGET FRIENDLY 'ACTIVE' RIBBON MICS which do not need an external preamp. Here is a reputable, budget friendly active ribbon microphone, the Golden Age R1 Mk3 (active), for about $300, see goldenageproject.com/microphones/ribbon-microphones/r1-active-mkiii/:

Here is a review, by Curtis Judd, see youtu.be/haZzCqgQZ50.

 AMBISONIC MICROPHONES

AMBISONIC MICROPHONES (LESS WELL KNOWN) ARE FOR RECORDING ATMOSPHERIC MATERIAL, like nature, but also to get a surround effect by recording in four directions; being four channels, which usually is for the purpose of 360 degree video, VR, and for gaming, and sport action. There could be a future here for recording Dolby Atmos height tracks – up to your imagination:

The microphone above is the Condenser Rode SoundField NT-SF1 Ambisonic Microphone, see rode.com/en/microphones/360-ambisonic/nt-sf1. Hear some of the possibilities at the Ambisonic Sound Library, see library.soundfield.com/. For recording with an Ambisonic mic, you need four inputs, which could be a digital interface like the Tascam US-4x4HR, see tascam.com/us/product/us-4x4hr/top.

For beginners, for easily making action sport or travel/nature recordings, you could experiment with the cheaper Zoom H3-VR, see zoomcorp.com/en/us/handheld-recorders/handheld-recorders/h3-vr-360-audio-recorder/. And, then, for DAW plugins on a quad track, the Waves 360° Ambisonics Tools, see https://youtu.be/pE_4YtbAEDA.

 ## BINAURAL HEADPHONES

BINAURAL HEADPHONES RECORD AS IF HEARD FROM EACH EAR, and are perhaps used more commonly for virtual reality (VR) purposes and gaming recordings, or for meditative music and sounds. The recording should only be listened to with headphones. Binaural recordings improve the immersive headphone experience.

Listen to these examples with a headset, from Ponds, at pond5.com/royalty-free-music/tag/binaural/. Here are some binaural microphones from 3DIO, FS Series, 3diosound.com/collections/fs-series-microphones:

Basically, if you record a sound off to one side, or in movement to or from one side, the sound becomes different as it reaches each ear, both in frequency and time of arrival. Binaural headphones will capture these nuances.

THIS SYSTEM DOES NOT WORK ON HI-FI SPEAKERS because in that case you are hearing a left and right channel in both ears, which would put binaural playback out of whack. Instead, for standard editing, panning to the left and right channels create a better result.

DUAL MICROPHONES

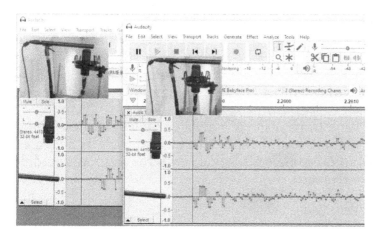

HAVE YOU WONDERED IF THERE IS A WAY TO GET THE BEST OF BOTH WORLDS, A CONDENSER MIC AND A RIBBON MIC; accuracy and tone. Sure.

I record vocals with a ribbon mic, a <u>Sure KSM 313</u> (plugged through a Grace M101), with a parallel shotgun condenser mic, a <u>Synco Mic-D2</u>. For online sessions, I set the distance at about a foot from my mouth, so they are out of webcam view. The mic heads are close but not touching.

For two mics, as above, the pickups of the mics need to be **time aligned**. It is worth experimenting with test placement, recording in stereo, and making a tap-sound recording and looking at the two tracks, zoomed in. Shown above, I hit two pens together for a sharp sound, then zoomed in to check the alignment. In the first recording (image behind), I presumed the end of the shotgun should be placed tip to tip. I was wrong, the shotgun mic (which is on track two) is delayed by about 0.5 of a m/sec. I redid this test until I got alignment, as can be seen in the overlaid image:

TIMING ALIGNMENT is fixing an issue that is similar to room echo where sound arrives at slightly different times.

Also, just to notice the difference between the condenser shotgun and the ribbon recording, you can see that the signals on the two tracks are slightly different - exactly - the shotgun mic is picking up more accurate information, while the ribbon is picking up more tonal information. In a DAW, after recording, you can sum the L and R channels to mono, then edit.

Here is a video on setting up two quite different mics, but in this case at an electric guitar amp cabinet, by Creative Sound Lab, see https://www.youtube.com/watch?v=VWc0HsQhpDU.

If you are recording into a DAW you might be recording many instruments each to its own track, and when doing so the mic distances may vary. In your DAW software you would then want to check time alignment of the different instruments. In this case it can be done, in post (after recording), with the Azimuth module of Izotope RX Advanced.

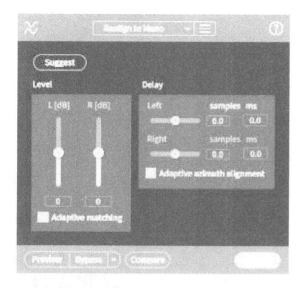

This whole timing alignment issue is also applicable if you are, for example, recordings a guitar while singing at the same time, using two microphones.

Another reputable option for automatic microphone alignment is Soundradix Auto-Align plugin, see soundradix.com/products/auto-align/#features.

As an aside, if you are remembering the chapter on Produced Sound, you will remember that if you want an instrument to appear further back in the soundstage, then a little reverb and/or echo, or delay, will achieve that.

As mentioned, the stereo file can be turned into mono. I do that in Izotope RX, by selecting Mixing > Mix to Mono:

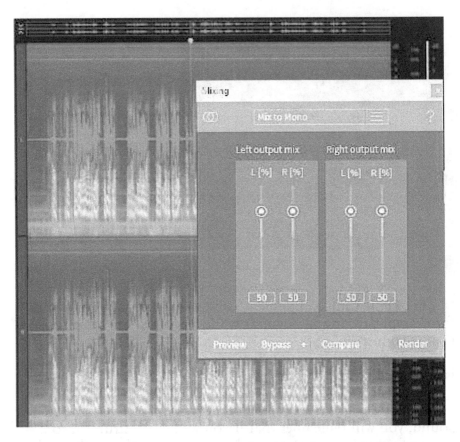

This makes both channels the same. One of the channels can then be pasted into a new mono track, or output to a 24-bit Wav file, with a bit of Dither (noise shaping).:

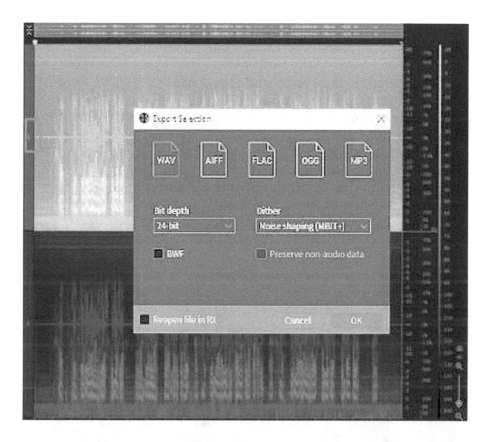

The **Townsend Labs Sphere L22**

IF YOU WANT ONE MICROPHONE THAT WILL DO EVERYTHING, here is a no gimmick cost saving approach; though it costs you to save it. It uses a dual capsule mic, into a stereo track recording, whereby you can closely emulate many mics. This is the **Townsend Labs Sphere L22**, see <u>townsendlabs.com/products/sphere-l22/</u>. The software plugin handles the mic emulation on the two channels. There are two microphone packages; the UA Sphere LX with 20 mic models, and the higher spec UA Sphere DLX with 38 mic models.

One reason to do this is to get the right microphone sound for your voice and instruments. As mentioned above, in professional recording studios the audio engineers match microphones to the talent's voice and genre. With this system you can experiment yourself, and, if you like, create double mic nuanced tonal variations.

The emulation choices expand if you have an Apollo Twin Mk II digital interface from Universal Audio. In that case, here are some links to emulations beyond what is in the initial package: firstly, the Bill Putnam Mic Collection, see uaudio.com/uad-plugins/mic-modeling/bill-putnam-mic-collection.html, and, secondly, the Ocean Way collection, see uaudio.com/uad-plugins/mic-modeling/ocean-way-microphone-collection.html.

Here is a review of the L22 modeling system, by Podcastage, see youtube.com/watch?v=kGD4H-XkARE, and note that because of the L22 full sound field pick up, you will need an acoustically treated recording space.

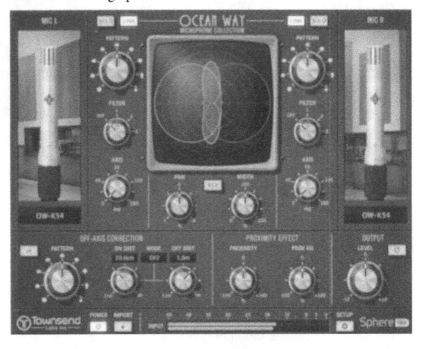

Yes, there are other software emulation plugins, but this gets the best reviews.

TUBE MICROPHONES

TUBE MICROPHONES USE INTERNAL VALVES for something akin to ribbon microphones, but with more of a warm vintage tone. They can be used on anything: vocals, guitars, drums, piano, strings, brass, woodwinds, or any other instrument. Think, *"vintage tone: warm and smooth to clean and bright"*. Have a look at the Warm Audio WA-251, costing about $2,000, see underline warmaudio.com/wa-251/:

And this review of tube mics, by mynewmicrophone.com, see mynewmicrophone.com/top-best-tube-condenser-microphones-on-the-market/.

These tube mics can cost a lot, so, another way to get that classic tube sound is to try out **tube plugins** which can emulate a tube sound, which you can use in your Digital Audio Workstation software (DAW) or media player. They add back an extra analog tone.

As an aside, for playback, you can also buy tube amplifiers, for a tube tone (see the chapter on 'Headphones, DACs & AMPS').

 ## SOFTWARE PLUGINS (EMPHASIS ON TUBE SOUND)

AS AN EXAMPLE, FOR PLUGINS THAT ADD A TUBE SOUND, there is the Abbey Road RS124 Compressor plugin, see waves.com/plugins/abbey-road-rs124-compressor, think *'Abbey Road, Beetles'*. If you record with a condenser microphone, then you can use a plugin like this in your DAW (Digital Audio Workstation) to get the feel of a classic tube tone on your own voice. If that works for your voice, you might think about getting a hardware tube mic (as above).

CHANNEL STRIPS, as discussed at the head of this chapter, used to be hardware, and are devices that group effects, mainly Compression and EQ. Compression brings the high and low volumes closer, so it is easy to hear everything in audio playback, such as in a car with background noise. Note, here, that compression or any effect does apply a certain amount of its own noise through the process of applying the effect. If you are aiming content for the audiophile or hi-fi market, you would be better off being very light on compression, because listeners will be in a quite room when listening, and therefore would prefer a truer to life dynamic range would be more realistic. The other main effect, Equalization (EQ) reduces negative room acoustics and shapes tone as a musical instrument preference.

* Channel strips (software based) or aspects of them, can be used on the input path in the digital interface (DI), specifically through the DSP features in quality DIs (see section on 'Digital Interfaces', below), and see the chapter on 'Equalization'. Applying channel strip type effects on the input path is useful if you are playing live, or want to apply a first step of 'gentle effects' to your recording, so that your post recording work is less drastic, even unnecessary, and also avoiding the risk of creating artifacts from applying too much EQ and compression effects at any one time (iterations of small effect applications create fewer artifacts than single substantial effects applications).

HERE ARE SOME CHANNEL PLUGINS that includes tube saturation options:
- Izotope Neutron Elements [The exciter module applies tube saturation], see izotope.com/en/products/neutron.html.
- TBProAudio's CS-5501 [The saturation module applies tube saturation], see tbproaudio.de/products/cs-5501.

IF YOU HAVE AN **APOLLO TWIN MKII** DIGITAL INTERFACE, (shown lower down in this chapter), there are some quality tube based channel strip VST plugins that will work on any Universal Audio DI. Their technology enables using the plugins in the digital signal processing path of the digital interface, with zero latency[*]. Some of these plugin example are:

- *Tube preamp only*: Manley® tube mic preamp, VST plugin, about $149, see www.uaudio.com/uad-plugins/channel-strips/manley-tube-preamp.html.

- *Tube preamp in a channel strip for vocals*: Manley's VOXBOX channel strip, VST plugin, all-tube vocal processor, about $300, see www.uaudio.com/uad-plugins/channel-strips/manley-voxbox.html.

- *Tube channel strip for voiceover, vocals and modern music*: The Avalon VT-737 Tube Channel Strip VST plugin, about $300, see www.uaudio.com/uad-plugins/channel-strips/avalon-vt-737-tube-channel-strip.html:

The hardware version, Avalon VT-737 Tube Channel Strip, can be bought at Sweetwater, for about $3,900, see www.sweetwater.com/store/detail/VT737SP–avalon-vt-737sp-class-a-mono-tube-channel-strip. Hardware gear may still be useful. See the table below for discussion on Pure DSD recording, where the hardware version would work well in the recording path.

* Zero latency is important if you are monitoring your own recording, so there is is no delay between the recorded sound and what you hear in your headphones. If you want to apply multiple plugins, however, the Apollo interfaces may still struggle to provide zero latency, in which case Universal Audio offers accelerators to overcome the delay issue, see www.uaudio.com/uad-accelerators.html. Apart from live recording, the preference is usually to apply plugins in post

(after you have recorded it) as part of your editing, mixing and mastering sessions.

I have experimented with the Tube-Tech Classic Channel mk II, from Softube, as a tube based VST channel strip, for about $230. Top left, is the preset section, which has default settings based on the type of music you are producing, see www.softube.com/complete-collection.

However, I did not end up using this in my DAW, because I record with a ribbon mic and I was happy with the tonal sound it provided. As an aside, there is another useful purpose. I do use this as a VST plugin for playback on my hi-fi system, placing it as a VST plugin in JRiver Media Player's DSP Studio, usually on the preset, 'JC Hi-Fi Vocal Chain'.

HERE ARE SOME OTHER TUBE-BASED EFFECT PLUGINS, depending on the effect wanted, listed by preference, by Slade Templeton, at sonicscoop, see sonicscoop.com/top-5-tube-saturation-plugins-showdown-best-dedicated-tube-saturators/.

OF COURSE, CHANNEL STRIPS DO NOT NEED TO BE TUBE-BASED. Here is an excellent example, the Steinberg RND Portico 5033/5043 Plug-in Bundle, about $400, see www.sweetwater.com/store/detail/RNDPlugBun–steinberg-rnd-portico-5033-5043-plug-in-bundle, which is modeled on the more modern transformer technology. I just stayed with the tube effects as a theme, leading of from discussion of the tube microphone as one type that provides a classic, *"arm and smooth to clean and bright"* tone.

Here are some other non tube based plugins, which I have found to be very effective, and might also be preferred for electrical music applications where a punchy tone is often preferred:
- **Fabfilter** plugins have long been praised as efficient and clean; applying the effects without changing tone, see fabfilter.com/. As an aside, I also use this as a VST plugin for headphone EQ on my hi-fi system, placing it as a VST plugin in JRiver Media Player's DSP Studio.
- **Gullfoss** uses EQ based on an audio perception model, for clarity, detail, and spatiality, see soundtheory.com/home.
- **Smart: EQ 3** has auto-eq settings as profiles, which can then be tweaked slightly, making EQing easier, see sonible.com/smarteq3/.
- **WaveriderTG** alters volume like a compressor, geared for dialog but does not change the tone, like a compressor does, see quietart.co.nz/waveridertg/.
- **SA2Dialog Processor** uses EQ, for dialog tonal clarity, used in the movie industry to bring clarity, see mcdsp.com/plugin-index/sa-2/. As an aside. I also use this as a VST plugin for movie playback on my hi-fi system, placing it as a VST plugin in JRiver Media Player's DSP Studio.

IF YOU ARE RECORDING MULTIPLE TRACKS, you will likely apply EQ and compressor plugins, to each track, perhaps using different plugins depending on the instrument recorded. Then, if preparing for stereo, you will want to balance the tracks, and do some EQ sculpting so each track can shine through in the mix.

THE FINAL STAGE OF APPLYING EFFECTS IS CALLED **MASTERING.** This, in part, is more so leveling everything to suit playback requirements of a particular genre or media, One well known package for that last stage is Izotope's mastering plugin integrated suite, Ozone. See this introduction by macProVideoDotCom, see youtu.be/YrH1a1RcnOY, and www.izotope.com/en/shop/mastering-software-and-plug-ins.html.

 FINDING THE BEST RECORDING POSITION

REF: *Image from https://www.thebroadcastbridge.com/content/entry/13629/the-art-of-microphone-placement*

FROM THE ARTICLE BY FRANK BEACHAM ON THE ART OF MICROPHONE PLACEMENT, see https://www.thebroadcastbridge.com/content/entry/13629/the-art-of-microphone-placement:

"…The player and the instrument contribute at least 50 percent to the overall sound. The room contributes another 20 percent to the overall sound, while **the position of the microphone contributes another 20 percent to the overall sound.** *Mic placement is the acoustic EQ and is responsible for the instrument's blend in the track. Microphone choice contributes another 10 percent to the overall sound…"*

Let us consider how to find a good recording position in a room for recording acoustic instruments, vocals or drums.

If you have done a good job of applying acoustic treatment in a recording space, then, all else considered, you will have clarity and crispness when you record. But what about the recording position? For listening, a common guide is 38% from the front wall (the wall behind the speakers), on the center line of the room. To record, turn around at the listening position, so you are about 38% away from the wall and step back so the microphone is in that 38% position [two people recording; two people with microphones 38% from a wall they have their back to, etc]. This is ball park. For recording, off center is probably better because there is usually a room mode that goes down the center of your room. So move the microphone to the left or right by about a foot. It will also pay to be at an angle [not square on to the room], so the projection of sound from you or your instrument is not reflecting straight back to you, but off on an angle. This gets us part way there.

The next aim is to establish the best recording sweet spot or spots, even more specifically than the 38% position. We know that when we measure EQ, sound alters in a room even by moving inches. You can listen for a best sweet spot, where the notes are as clear as can be, and you can use an App. For listening on your own try covering one ear while cupping the other. For an App, try the TUNIC Unisono App, a sweet spot locater, which shows a circular tonal image when the tone of any particular note is pure. By testing it with the most common notes, or simply play, say or sing a sample of what you want to record you can locate a place or places where you will get the purist tonal recording; the most circular images on the TUNIC Unisono App.

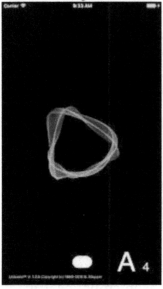

Check these links out:
youtu.be/0a97GEf4MzQ
youtu.be/5fn5GccEpX0
appadvice.com/app/unisono-sweet-spot-indicator/1136709641
Last I checked, you can register and try out the APP for free, for two weeks. Sure, not every note you play will be spot on in one position, but you can aim for the most common ones.

PORTABLE VOCAL BOOTH

ONE THING TO MENTION, THOUGH NOT SO MUCH TO RECOMMEND, IS A PORTABLE VOCAL BOOTH, designed to reduce room mode interference, see this article by Sam Inglis, on Sound on Sound, which tests some of the main brands at the Acoustics Laboratory at the University of Salford, soundonsound.com/reviews/how-effective-are-portable-vocal-booths. The tests were in a room with a reasonable amount of acoustic treatment. Unfortunately, although they do remove some room reflections, they also introduce reflections of their own. Therefore, I would suggest you rather focus on improving your room acoustics. In the interim, if you do not yet have your acoustic treatment worked out, or want to face a window when you record, a portable vocal booth could be an effective stand in.

I will now move to talking about digital interfaces, since a microphone usually connects to one.

DIGITAL (AUDIO) INTERFACES

A GOOD WAY TO UNDERSTAND A DIGITAL INTERFACE, is not just that it turns analogue into digital audio, but that at a more basic level it has one essential purpose - to replace your shitty computer soundcard. Okay, it may not be completely shitty, but comparatively it is. The same issue applies to AVRs, which are designed to take over from comparatively shitty TV audio soundcards. You can listen to a TV as it comes out of the box, but if you step around the TVs soundcard with an AVR or Soundbar (which also have their own soundcards), you will get a far better audio experience.

I also want to mention that while Digital Interfaces (DIs) are designed for recording audio, and for playing back audio while editing and mastering, they can of course be used as a DAC before a Power Amp for a lounge listening situation. The main difference is that a DI includes an Analog to Digital Converter for digitizing the microphone signals. To prove the point, I have used my BabyFace Pro DI in a lounge listening environment, playing to six speakers (2 powered speakers, and out to two Power Amps each to a pair of stereo speakers). More on this will be covered in the chapter on 'Headphones, DACS and AMPS'. Now, back to our focus on Digital Interfaces for audio production purposes.

IT IS FAR BETTER TO HAVE A QUALITY DIGITAL INTERFACE (DI), not an entry level one, for a number of reasons:
1. The internal Analogue-to-Digital Converter (ADC) driver/s need to be of high quality, so as not to introduce noise into the digitized audio.
2. The quality of the internal pre-amplifiers (for boosting the microphone signal) will make it possible to turn the recording gain/volume right up without introducing hiss (likely necessary for Dynamic microphones).
3. The internal Digital-to-Analogue Converter (DAC) needs to be of high quality to handle any set of headphones (no matter the ohm Ω load).

4. It is better to have two headphone jacks, so one can be used for a headphone, and the other, as an option, for a second set of headphones, or as a preamp to a hi-fi amplifier, as I have done, as mentioned above.

5. Better DIs have DSP (digital signal processing) with subtractive EQ, so if you like you can adjust frequencies to counter the worst room modes (which negates or reduces the amount of EQ you need to add to post recorded audio).

6. An independent power connector (not just power from the computer via USB) ensures the loads above are handled easily, rather than drawn from your computer power processing.

HERE ARE SOME GOOD DIGITAL INTERFACES (Windows and Mac compatible), which meet the above criteria. I have included some links to YouTube videos of the controller apps so you can see the range of settings:

Digital Interface	controller App	Useful video about the controller Apps (DSP focus)
MOTU UltraLite MK5	Cue Mix 5	youtu.be/ aSWbuzE8Unc
TASCAM Series 208i	Tasman Control	youtu.be/ IF9kqn5hfGI?t=512
Steinberg UR44C USB	dspMixFx	youtu.be/ eqW5hwFSqkA
BabyFace Pro FS	TotalMix FX	youtu.be/ yfhNh4WmPso
Apollo Twin USB or Apollo Twin MKII for a Mac with its thunderbolt 3 port	Console App	youtu.be/ Axcvj5IvMpY
For higher audiophile quality recordings, where you are very concerned to record in the best analog format, you can move		

away from PCM to the DSD format, as discussed here by Paul McGowan of PS Audio, youtu.be/W98Bs5t9oec. For this purpose, I have provided information below on the Merging+ Anubis Premium digital interface, which can record up to 352.8 kHz (DXD), and up to DSD256. There is no worthwhile advantage of recording above that. And, yes, the cost goes up.

Merging+Anubis Premium, about $2,500, can record up to 352.8 kHz (DXD), and up to DSD256. Here is a review /introduction, by Streaky, see youtu.be/jggg-9 TBM4. See the product info at merging.com/products/interfaces/mergi ng+anubis. Among other places, it is sold here, https://vintageking.com/merging-technologies-anubis-premium-sps

For a DSD capable DAW, use Merging's Pyramix Pro version, which includes video support, core plugins, surround sound panning, and loudness metering, for about $600, such as from Vintage King, vintageking.com/merging-technologies-pyramix-native-pro.

You can physically connect the Anubis to your computer with a USB 3.0 to Gigabit Ethernet NIC Network Adapter, see startech.com/en-nz/networking-io/usb3 1000s.

You can achieve a **multi-channel or Atmos hardware recording** and playback setup, with the cost going up, by adding a Merging+Hapi

to the Merging+Anubis Premium, inclusive of an ADA8P Mic/Line Input and Output card,

adding up as $2,500+ $4,500, (Anubis x 4 analog outputs & Hapi x 8 outputs = total 12 outputs). Among other places, they are sold at Vintageking:
-
https://vintageking.com/merging-technologies-hapi-mkii

The controller App is on the device, via a viewing window, see youtu.be/4ilG22qpizY.

(a) Record as you would normally, up to a 352.8 kHz sampling rate.

(b) Record in PCM 352.8 kHz (DXD), which allows for editing and mastering in Pyramix Pro, then output to PCM 192kHz and/or DSD256 to upload to websites offering those formats (downsample for other services).

(b) Record as a 1-bit DSD256 session. In that format, you cannot do any editing or mastering, so make multiple recordings until you get it just right. Then, export direct to an output file. Some audiophiles prefer this as a purer analog-like rendition. Pay extra attention to microphone placement, perhaps also use a hardware channel strip in the recording input path, so as to manage EQ and compression with tube technology. I have listed some below*. If you are recording DSD in a well acoustically treated room, you can get by without EQ and compression. EQ is not needed if a room is acoustically treated with a second step of room measurement and targeted bass trap building. Compression is not

-
https://vintageking.com/merging-technologies-ada8p

See the Merging basic Atmos setup diagram, merging.com/assets/img/merging/products/anubis/2021/diagrams/resized/Atmos-7.1.4-SPS-2.png.

You may need to upgrade your computer to a high spec gaming computer.

For multichannel editing, mixing and mastering, you would need the Pyramix Premium DAW. Here is some information on how Pyramix meshes with the Dolby Atmos Renderer, see merging.com/products/pyramix/key-features.

You can physically connect the Hapi with microphones and to speakers, with a DB-25 connector to needed when the end user listeners have a good quiet listening environment.

If you supply the DSD or DXD master MTFF (Merging Technologies File Format) to nativeDSD they will produce all the DSD bitrate deliverables, including an upscaled DSD512, for sale.

In this case it is a bit tricky to use it as a DAC for HI-Fi listening, but you can, see https://audiophilestyle.com/ca/reviews/merging-technologies-anubis-on-the-desktop-r1084/. P.S., the Qobuz app works directly with the Merging asio drivers.

XLR male or female, see proaudiola.com/categ ory-s/429.htm,

or DB-25 connector to RCA, see proaudiola.com/categ ory-s/421.htm.

* Having a hardware tube based channel strip in the recording path may be advantageous when recording in DSD, with tube based flavor for any instrument and voice, inserted between the microphone and processor, such as the Merging+Anubis above. Being in the recording path is pertinent for pure DSD recording, whereby there is no editing or mastering possible if you keep the signal pure DSD and output directly to a DSD file (being an audiophile preference). If however, on the Merging+Anubis you record in DXD you can use the Anubis EQ and Compression, as well as in the DAW, and then you can output as DSD. That is not as pure as DSD though). These hardware units will handle any instrument, including vocals:
- Universal Audio LA-610 Mk II Tube Channel Strip, about $2,000,

see www.sweetwater.com/store/detail/LA610mk2–universal-audio-la-610-mk-ii.
- Retro Instruments Retro Powerstrip Tube Channel Strip, for about $3,500,

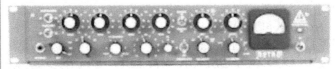

see www.sweetwater.com/store/detail/RetroPowerstrip–retro-instruments-retro-powerstrip.
- Avalon VT-737sp Tube Channel Strip, for about $4,000,

see www.sweetwater.com/store/detail/VT737SP–avalon-vt-737sp-class-a-mono-tube-channel-strip.
- Manley VOXBOX Tube Channel Strip, for about $5,500,

see www.sweetwater.com/store/detail/VoxBox–manley-voxbox-channel-strip. It also has a DeEsser.
Here are some others, a couple being more budget friendly, recommended by Max McAllister, of Audio Engineering, see producelikeapro.com/blog/best-tube-preamp.

DIGITAL AUDIO WORKSTATION (DAW) SOFTWARE

IF YOU ARE GOING TO RECORD AUDIO, EDIT, MIX, AND MASTER, you are going to need a Digital Audio Workstation (DAW). Here is a list of popular DAWs, showing the main differences, by the DAW Studio, see thedawstudio.com/resources/daw-comparison-chart/.

I HAVE BEEN USING STUDIO ONE, see presonus.com/products/Studio-One, because my work has been more in audio narration and there is a good Facebook group on working in Studio One for that purpose, the Don Baarn's Studio One Narrators, VO & Podcasters, see m.facebook.com/groups/StudioOneNarrationVO/.

WHICH DAW IS BEST FOR YOU? Before committing to one, it will pay to search the internet for forums and user groups where other people are doing the same or similar projects as you are, or want to, then see which DAW or DAWs they are using. That way there will be much more specific help available, and you can connect in to that community to help keep you motivated. I mentioned, in my case that I use Studio One, and in the same breath that I connected up with a Facebook group working with Studio One for the same purpose.

THEN BECAUSE EACH DAW DOES TRY TO DO EVERYTHING, you need to work out the minimal things you need to know to control it to do what you want to do. I do not think I need to go into more detail in this book, as there are many YouTube videos introducing and going into detail of each DAW. Because DAWs are so comprehensive it can be a bit of a challenge to work them out. The antidote is to locate a forum or Facebook group doing the same thing as you, and focus on your area, using the DAW you have chosen. Perhaps more useful, will be a look at how mixing and mastering in Atmos may be accomplished.

SPATIAL SOUND: UP-MIXING TO ATMOS IN A DAW

ONE SPECIALIST AREA, PERHAPS ON A GROWTH CURVE, IS MIXING AND MASTERING MUSIC FOR DOLBY ATMOS PLAYBACK. While some may be into the movie editing scene, I will focus on creating Atmos music, which at any rate would be a part of embedding music into a movie mix. One reason this may be catching on is that a number of AVRs now have class AB amplification, which generally sound more musical than the cheaper class D amplifiers; some better Soundbars also have class AB amplification. This means that listening to streamed music on an AVR or Soundbar setups will likely be increasingly popular, and so an interest in Atmos-abled music.

Let us first consider what Atmos is in relation to spacial sound. To discuss it, we will jump to how Atmos content is played out to a listener, and so to cover surround sound, up-mixing, down-mixing, and how to work with the Atmos format.

Dolby Atmos takes the bed of surround sound channels, such as in a 5.1 system (5 speakers: Left, Center, Right, Rear left, Rear right, + Subwoofer and **adds multiple 3D moving objects** which play through height speakers in a home entertainment system, thus extending a surround sound system with moving 3D special effects. Here is an engineer from Adam Audio explaining the basics of how Dolby Atmos builds on surround sound, youtu.be/jyOOxI8-zSQ.

In case you have the question, height speakers are simply any pair or pairs of passive speakers, aimed above our heads, and are assigned for the playing of Atmos channels by your AVR or Soundbar. Some speakers are designed, more so, with that purpose in mind.

Up-mixing, in terms of what you might want to do in a DAW, typically involves taking a master project file and up-sampling/turning it into tracks applied to surround channels, and adding movable 3D objects. For the 3D objects, there may be a track or two of higher frequency musical effects that can be reapplied as 3D objects. The Penteo Pro plugin, see perfectsurround.com/, can handle the upsample to as many channels as 9.1.6. Here is a look at how this can be done with the plugin, Penteo 16 Pro, see youtu.be/6AO48_B5GIA.

A WORD ABOUT DOWN-MIXING, which is what can happen on the end users listening system. Down-mixing involves turning multiple channel audio into fewer channels. For example, if a Dolby Atmos file has 9.1.6 channels, and your home entertainment system was set up for 5.1.2 your Audio Video Receiver (AVR) will down-mix it on the fly, so you still get all the sound, but reshaped for your speaker system. If you play a Dolby Atmos file from the Internet as streamed content, for example, you will only hear it as maximally intended if you have Dolby Atmos compatible connections and playback devices. If they are not present, let's call that a "playback bottleneck", you will get a default down-mix happening.

The 3D Atmos effect channels can be heard in a DAW with the help of the Adobe Atmos renderer, see professional.dolby.com/product/dolby-atmos-content-creation/dolby-atmos-renderer/. Then the exported mp4 can be distributed, via a service, like Avid's Avidplay, for onloading to services such as Amazon Music, Apple Music, TIDAL, in fact over 150 streaming services, see avid.com/press-center/avidplay-subscription-service-now-enables-indie-artists-and-labels-to-distribute-tracks-mixed-in-dolby-atmos-for-apple-musics-spatial-audio/.

Here is a video where editing and mastering for Atmos is discussed, by Home Theater Geeks, see youtu.be/v8Xto_1AjQs.

The easiest way is to do the **mixing is on headphones**, thus saving on the cost up upgrading your DI and monitors which you would need to do if you want an in-room listening reference. Here is how to mix Dolby Atmos with headphones, by Dolby Institute, see youtu.be/qs1tffnAjPE.

TO PERIODICALLY CHECK YOUR WORK ON A DOLBY ATMOS ENTERTAINMENT SYSTEM, you could play your exported mp4 files with the free VLC Player, see vlc-media-player.en.softonic.com/, by selecting Interface settings > Audio > HDMI/SPIDIF audio Passthrough> Enabled. Then Device > selecting your AVR, which will be in the drop down list if connected up. For Mac, Frank Martin explains how, youtu.be/GUblp6EVtJ0.

IF YOU WANT TO MIX WITH MONITORS/SPEAKERS, there is a bit of setup and cost involved. The best setup I can suggest has already been introduced above, being the Merging+Anubis Premium, with a connected Merging+Hapi device (with an ADA8P Mic/Line Input and Output card). This setup, as mentioned, can work in DXD and DSD formats. Multiple speakers can also be configured with this system, with EQ (during editing) handled with Sonarworks Merging+Anubis SoundID Reference plugin, see sonarworks.com/soundid-reference/integrations/merging-anubis.

Here is an alternative option. You could upgrade to an Apollo 16, see uaudio.com/audio-interfaces/apollo-x16.html, setting you back about USD3,500, plus the cost of adding more monitors onto any existing studio monitor setup. Like the Merging +Anubis Premium solution, you can manage EQ room correction in the Console App. For example, you could get the UAD plugin Cambridge EQ, see uaudio.com/uad-plugins/equalizers/cambridge-eq.html, and insert it into each of the Console App channels. Then Measure your room EQ to each monitor/speaker, and enter up to 5 EQ settings in the applicable channels (see the Equalization chapter for other ways to add room calibration EQ settings). The Apollo option is cheaper than the Merging one, but does not have the DSX, DSD versatility.

AND WHAT DAW SHOULD YOU USE FOR ATMOS? Let me start from the Penteo plugin's' compatibility list. Checking the requirements at the bottom of the page at perfectsurround.com/, the list, as of checking (Dec 2022), includes: ProTools, Apple's Logic Pro, Reaper, Steinberg's Cubebase Pro, Steinberg's Nuendo, Pyramix, Resolve, and Ableton Live. You would use Pryamix if your were setting up the Merging option and wanted to work with DSX and/or DSD. Otherwise, Pro Tools and Apple's Logic Pro seem to be ahead of the pack at the moment, with respect to Dolby Atmos. For example, here are some just out (Dec, 2022) reading resources on how to mix in Atmos with Pro Tools, see dingdingmusic.com/-titles-/dolby-atmos-2.html, and Atmos with Logic Pro, see dingdingmusic.com/-titles-/dolby-atmos-3.html. Also, you will notice at Dolby Atmos Learning, see professionalsupport.dolby.com/s/learning, the training is more so geared to Pro Tools and Apple's Logic Pro. However, I think the other DAWs will be catching up, fast.

See more information about Dolby Atmos for the end user in the chapter on 'Streaming and Play-through'.

SUMMARY

IN THIS CHAPTER we considered what traditional analog dependent channel strips do in the recording pathway, and nowadays as settings in a digital interface and/or a series of plugins.

We saw how we can choose plugins to emphasize an analog tube based tone – not that we have to do it, but to show how you can veer in an analog direction if you wish to do so. We also covered different microphone types and their general uses, even a dual microphone setup – in a later chapter I will cover how some specific models are more suitable to different genres of music. We also needed to know how to find a recording position in a room. Then, since we need a digital interface to record, those were discussed, along with the various DSP features, and, I introduced, along with a range of digital interfaces, one that can be used for the nearest-thing-to-analog native DSD level and Atmos recordings.

There were a few more aspects to cover: the various connections in and out of digital interfaces, and an introduction to Digital Audio Workstation software (DAW).

Nearing the end of this chapter, I linked back to Dolby Atmos, by way of discussion on up-mixing to Atmos in a DAW. The theme of Atmos is picked up in the chapter on 'Streaming and Playthrough'.

CHAPTER 7
HEADPHONES, DAC, AMPS, AND TV-AVR INTEGRATION

"The language of translation ought never to attract attention to itself", by John Hookham Frere.

I START THIS CHAPTER LOOKING AT HEADPHONES, THEN MOVE THROUGH ALL THE DEVISES IN A HI-FI SYSTEM, AND SEE HOW THAT INTEGRATES WITH A TV/AVR/SURROUND SOUND SETUP. Here is an initial integrated system, which for Hi-Fi relies on a Class AB AVR and its power amps for Stereo and Surround sound processing. HiFi is processed through a DAC with the computer being the streaming device. The Smart 4K TV is also used as an extended computer monitor with the TV Input being switched for watching movies etc via the AVR:

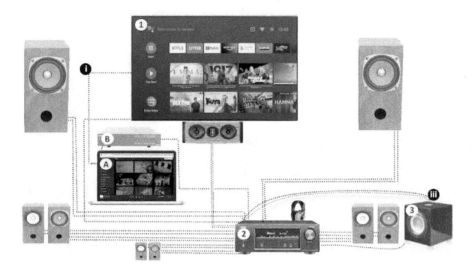

Music shaped by parallel system to AVR; sharing the main Left and Right Speakers. Here is an expanded Hi-Fi setup which utilizes a Tube Pre-Amplifier for its extra holigraphic and tone shaping ability, this on-connects to a Solid State Power Amp and Headphone Amp. A switch is used so only one set of main left and right speakers is needed, it switches between the Hi-Fi to main Left and Right speakers, and the Smart TV to AVR Surround Sound/Atmos system:

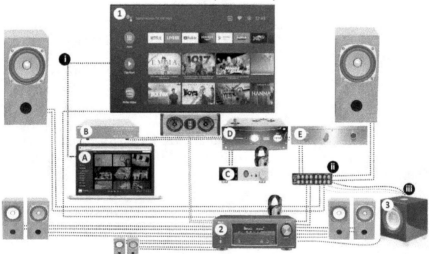

Above, I am integrating the system with a separate Tube Pre-amp and Power amp because I believe it provides the best sound quality, tone flexibility, and cost effectiveness. However, it would be remisss to not explain **the alternative** of using an Integrated AMP, which is a combination of (D+E+ii) a Pre-amp and Power amp, thus not requiring (ii) a 2-way Amp to speaker switch.

You can use an Integrated Amp to achieve the same sort of integration as above, as long as it has 'Main-in' connections on the back, and on the front a corrsponding 'Direct' / Home Theatre Bypass' / 'Power Amp-in' / or 'AV-in' selector - those are different naming conventions for the same thing. Here is an online list of Integrated amp with that feature, see https://www.audiophile.no/en/articles-tests-reviews/item/426-amplifiers-with-processor-input. From that list, here is the Onkyo A-9150, showing its Main-in connection and corresponding 'Direct' selector:

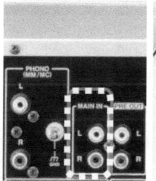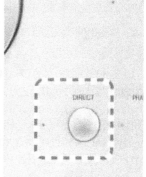

Equally, you could use a an Integrated Tube Amp that has the 'Main-in' connections and 'Direct' or equivalent switch, like the Musical Paradise MP501, see https://www.musicalparadise.ca/store/index.php?route=product/product&product_id=55.

From the AVR, you would go out of the AVRs Pre-amp RCA jacks for the main Left and Right channels [thus not using the power amp in the AVR for those channels], and on-connect into the Integrated Amp to the 'Main-in' L&R connectors [thus stepping around the Intgrated Amp's Pre-amp]. In that case, since the volume control of the Integrated Amp is part of its pre-amp, it should not work - you would control volume on the AVR.

Once connected up to the AVR, you listen to movies with the 'Main-in' selected. and you deselect it to play stereo music direct through a Streamer or Streamer then DAC.

AS YOU READ THROUGH THIS CHAPTER I will explain various gear alternatives, which can serve as a reference, even if you setup your gear differently.

SINCE HEADPHONES AND EARBUDS, are used for mobile applications, and a couple of the suggested mobile DACs can also double up as desktop DACs, I will start with them.

EARBUDS

EARBUDS COME IN ALL SHAPES AND SIZES, and if you pay a bit more, you pay a bit more, you will get a good snug fit in your ear. Another important feature is Bluetooth connectivity, because for practical reasons we often don't want wires hanging from our ears to our pocket, bag or desk. Bluetooth is something of a compromise if we do want lossless hi-Res music. Bluetooth uses compression and there is an ever so slight delay which can cause lip sync issues for movie viewing. The compression works in the same way mp3 works by removing the frequencies humans cannot hear. The difference will most likely not be noticeable when on the road, but it is a small compromise. With bluetooth enabled headphones, check your smartphone has aptX HD or LDAC (as explained above) for hi-Res compatibility.

Enhanced hearing:

IN TERMS OF HEARING ENHANCEMENT, we are talking of applications to accommodate hearing loss. In this case, a quality device is important because we are discussing streaming music to them. These may have Blue tooth for streaming music and a device-specific apps for controlling variable hearing situations. Take, for example, the Signia touch control app. Rated by tech.co, see tech.co/hearing-aids/android-hearing-aid-app as one of 2022's top hearing aid apps for Android. The app is designed to work with the Simiens hearing aid. Or, for iPhone, see the Jabra GN, voted the best here, by Seniorliving.org, see www.seniorliving.org/hearing-aids/best/iphone/, which comes with the Jabra Enhanced App:

Having an ear piece that is custom molded and/or is more or less invisible is an obvious advantage. Here is a list of the best invisible hearing aids by Soundly.com, see www.soundly.com/blog/best-invisible-hearing-aids. The price goes up, but in that review the best is the Starkley Genesis AI Custom, which you will notice is Bluetooth compatible, with an App which also incorporates Livio AI for real time language translation, and back-to-the-other person on-screen translation from you, see https://www.starkey.com/blog/articles/2018/10/Livio-AI-translation-tool. That feature is an integration of what portable AI language translators do.

  CLOSED-BACK HEADPHONES

CLOSED-BACK HEADPHONES ARE THE MOST COMMON TYPE sold in music stores. They typically boost the bass because users often prefer that for visceral skull impact. Only speakers, especially subwoofers, can provide a visceral body impact. I cannot see the point in bass boosted headphones, since if you first of all get quality headphones, the bass will be naturally good, and then if you wish, you can EQ them for a greater bassy effect to you liking (see the Equalization chapter).

Closed-back headphones generally cancel out external noise – sometimes you want that, but, there is a compromise in doing that. Resonances, like room modes and echo, are caused by the internal echo of a closed-ear cup, ending up as a sort of closed-in sound. The other weakness is that the sound pressure in a closed-back design pushes against your eardrum in an over-time fatiguing way. For quality closed-back headphones, expect to pay something in the range of $600, like this Focal Listen Professional closed-cup design, see focal.com/en/pro-audio/monitoring-speakers/professional-headphones/listen-professional).

Note: low impedance headphones (8-32 ohm) typically work

well with portable devices; high impedance headphones require more source voltage, such as provided from a device that needs to be plugged in. I will indicate the impedance of any headphones listed.

Here are some other reputable closed-back headphones:

Dynamic headphones (most common):
Airpods Max (closed)*[low 32 ohm impedance]*,
AKG K701 (closed)*[66 ohms]*,
Q701 (closed)*[62 ohms]*;
Audio-Technica ATH-M40x (closed)*[35 ohms]*,
ATH-M50x (closed)*[38 ohms]*,
ATH-M70x (closed)*[35 ohms]*,
Beyerdynamic DT770 (closed)*[80 ohms]*,
DT1770 Pro (closed)*[48 ohms]*,
Sennheiser HD 280 Pro (closed)*[64 ohms]*,
Sony MDR-7506 (Closed)*[64 ohms]*.

Planar headphones (more accurate):
Audeze LCD2 (closed)*[70 ohms]*,
Audeze LCD-2C (closed)*[70 ohms]*,
Audeze LCD-X (closed)*[20 ohms]*,
Audeze LCD-XC (closed)*[20 ohms, for mobile devices]*,
Audeze LCD-MX (closed)*[20 ohms, for mobile devices]*;
MrSpeakers Ether (closed)*[22 ohms, for mobile devices]*;

HAVING BLUETOOTH HEADPHONES, means you do not need to worry about cables, but it compromises slightly the audio signal because Bluetooth does use compression and there is an ever so slight delay which can cause lip sync issues for movie viewing. The compression works in the same way mp3 works by removing the frequencies humans cannot hear. The difference will most likely not be noticeable, but it is a small compromise. With bluetooth enabled headphones, check your smartphone has aptX HD or LDAC (as explained above) for hi-Res compatibility.

OPEN-BACK HEADPHONES

OPEN-BACK HEADPHONES (WITH AN OPEN CUP) WILL PROVIDE THE BEST SOUND reproduction because the open-back design minimizes resonance caused by the cup themselves. Because of this, this type of headphone is often preferred for working with audio, for editing, though not for monitoring. The open-cup lets out the back speaker diaphragm sound and pressure, so along with significantly reduced ear-cup echo, you will get a more spatially natural sound.

You will see many of these if you do a search for "Best studio headphones". Note that those that are closed-back are for monitoring while you are recording because the sound will not bleed out of the headphones back into the microphone, whereas open-backed headphones are for editing, mixing, and mastering audio, and for end user listening.

With open-back headphones, you might think other people can hear what you are listening to. That would be true on a bus, but at home or in an office, a person two meters away will not hear it if your volume is modest. You will be able to hear external noise quite clearly – which I think is good if you are working in an office, so other people can get your attention. Conversely, if other people are speaking on the phone, you will hear it.

Here are some quality open-back headphones:

Dynamic headphones (most common):
Audio-Technica ATH-R70x (open)*[470 ohms]*;
Beyerdynamic Pro DT990 Pro (open)*[267 ohms]*,
Beyerdynamic DT1990 Pro (open)*[80 ohms]*.
Sennheiser HD600 (open)*[300 ohms]*.

Planar headphones (more accurate):

Audeze LCD-1 (open)*[35 ohms]*,
Audeze LCD-2 (open)*[70 ohms]*,
Audeze LCD-3 (open)*[110 ohms]*,
Audeze LCD-4 (open)*[200 ohms]*,
Audeze LCD-4z (open)*[15 ohms]*,
Audeze LCD-X (open)*[20 ohms]*.

Electrostatic headphones (even more accurate), considered hi-end:

My reservations are similar to the use of pure silver in speaker or headphone cables, that you might hear defects, along with the additional clarity, not heard by audio engineers, who have used copper in the audio production process. That would not apply to Native DSD recorded audio in the DSD format. Electrostatic headphones require electrostatic headphone amps (refereed to as Earspeaker Driver units), so if you are looking at these headphones you are probably going to be looking at a bundle. A couple of the bundles seen at Sax https://staxheadphones.com/collections/stax-bundle-set are:

Bundle Set SR-L300 + SRM-400S (open), about $1,600
Bundle Set SR-L500MKII + SRM-500T (open)*[Tube based]*, about $2,100
Bundle Set SR-007MK2 + SRM-700S / SRM-700T (open), about $5,000
other:
Koss ESP 950 Electrostatic Headphones Kit (open), about $1,700.

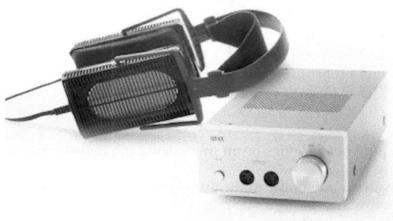

Connect it from your DAC.

 SEMI-OPEN HEADPHONES

SEMI-OPEN HEADPHONES (WITH A SEMI-OPEN CUP) WILL PROVIDE A COMPROMISE between a closed-cup and open-cup headphone, providing a semi-open surround sound feeling, which is less pressure on the ear drum, and so are a better design than closed-back headphones, if you can manage the small amount of sound leakage to the outside world. They could be used as a second headphone when editing, mixing and mastering, and they could be used for listening to music or movies, not on a bus, but in a library, office or elsewhere. I have the AKG K 240 *[600 ohms]*, see akg.com/Headphones/Professional%20Headphones/K240-Studio.html, presently being used by my son. Or better, Beyerdynamic T1 now 3rd generation *[600 ohms]*, if you can handle the cost, see global.beyerdynamic.com/t1.html.

 SPECIALIST ATMOS MULTI-DRIVER GAMING HEADPHONES

DOLBY ATMOS SOFTWARE CAN RECREATE ATMOS EFFECTS TO A STEREO TRACK FOR ANY HEADPHONES very efficiently, and while that can be achieved for any of the headsets above, it is worth mentioning one gaming headset (the **Razer Tiamats**), designed with multi-drivers (10 in total) to mimic a surround sound and Atmos experience. The Razer Tiamat gaming headphones, have five small speakers in each ear-cup, and seem the only model, as at Jan 2024, which mimics Atmos using actual multiple drivers. Perhaps this idea will catch on; perhaps it is unnecessary.

Other models, marketed as Atmos headsets, are in fact stereo headsets, but with additional software features – I don't see the point of purchasing one advertised in that way. For example, one I was looking at pointed out how the 40mm drivers provided an advantage for Atmos Playback, but, my Sennheiser HD600s, have 42mm drivers! At this stage, for lounge listening, I think one is far better off with a quality open-back headset, as above, and having compatible Atmos devices and software.

See more information on Atmos play-through in the section on Atmos for Headphones in the 'Streaming and Play-through' chapter.

 HEADPHONE MODIFICATIONS

A COMMON HEADPHONE MODIFICATION is to upgrade the headphone leads and connections, another is to purchase an upgrade kit. It is always better to improve headphones at the physical (analog) level rather than with software.

You can upgrade headphone leads on better quality headphones where you can disconnect the leads. However, with quality headphones, the original leads should be absolutely fine. Then, an upgrade is only useful if you damage the original lead, such as running over it with a castor chair.

If you really want to be tempted to upgrade your headphone leads, I suggest you look for pure copper, quality braided leads, and quality plugs, such as Neutrik jacks. Do not bother with pure silver wire or silver coated wire. Yes, silver is a better conductor than copper, but what you are listening to will have been produced with copper wires. Pure reproduction will come from pure copper.

WITH WELL KNOWN HEADPHONES THERE CAN BE OTHER MODIFICATIONS available. In my case I added the Custom Cans HD600 CNC Copper Mass loading mod DIY kit, see https://customcans.co.uk/shop/product/hd600-hd650-cnc-copper-mass-loading-mod-diy-kit-2/ to my Sennheiser HD600s. This brought noticeable clarity to the bass (better than tilting up the bass frequencies in EQ), and added smoothness to the upper mids and highs. Along with the EQ added (see the Equalization chapter), this achieves just that extra bit for, what to my ears is, a magnificently crisp and clear sound.

In general, check out reviews before adding a modification. And, if you are going to add EQ, you need to notch back any library EQ settings to compensate for the improvement of your modification.

 SPATIAL SOUND: DOLBY ATMOS FOR HEADPHONES

IF YOU ARE USING A PC, or SMARTPHONETO ACCESS SERVICES like Netflix, Amazon Prime, Netflix Music, Tidal, or Amazon Music, one option is to set up your Smartphone, Mac or PC to hear the Dolby Atmos sound experience on headphones.

From streaming services, you can search for Dolby Atmos playlists. Since these have been mastered in Atmos they will sound better than using an Atmos software setting over a stereo file.

Try this Surround test free download, inclusive of Dolby Atmos, from immersiveaudioalbum.com, see https://immersiveaudioalbum.com/sonic-brand/

Atmos control on a stereo device will render a stereo signal into a spatial audio Atmos rendition. It takes in the 7.1.4 channels of Atmos audio and virtualizes and binauralizes them using a process called HRTF (Head Related Transfer Function). The listener then perceives the sound as if around their head. It does this by modeling the time difference between sounds hitting either ear, and the acoustic shadowing caused by our pinnae (outer ear), head and torso. When you stream, if the music was mastered in Atmos there will be an Atmos and Stereo version; the one playing according to whether you have your Atmos settings to on or off.

Apple's newer products have an Atmos renderer built in (or rather they call it Apple Spatial Audio): MacBook Pro, iPad Pro, Apple TV 4K, MacBook Air, iPhone 12 pro, iPhone 12, iPhone 11, iPhone x, iPhone SE. For Android smartphones, there are: Samsung Galaxy S9 & 10, Note 9, Oppo Reno, Sony Xperia, OnePlus 7 Pro, Razer Phone. Many newer Smart TVs, Soundbars, and AVRs are Atmos compatible.

Let's look at setting up Dolby Atmos for Headphones on a Windows PC? First, remember the proviso, as proof of concept, now read on. Let's look at setting up Dolby Atmos for Headphones on a Windows PCs. First, you need the Dolby Atmos for Headphones software for Windows 10 or above, see apps.microsoft.com/store/detail/dolby-atmos-for-headphones/9M V4JDRVQTNV). It is less than USD20. Then check Spatial sound>Dolby Atmos for Headphones. On Windows, you will see the settings, with a right mouse click over your sound icon. Here Dolby Atmos for Headphones needs to be selected:

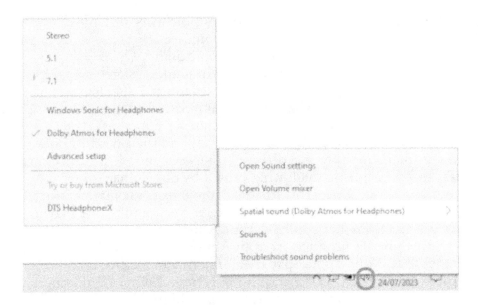

Any audio played on your PC goes through your Spatial sound settings before it goes to your selected audio driver, with the physical pathway out of you computer being either via the headphone jack, USB or HDMI. However, Windows OS Spatial sound settings are limited in two ways. Firstly, if you are playing hi-res audio, it downsamples Audio to 16-bit/48 kHz (DVD quality).

This downsampling limitation on PCs may not be of concern to many users, but, because of that, I suggest aiming for iOS devices, because Apple Spatial Audio adjusts automatically to use lossless 16-bit/48 kHz, 24-bit/48kHz, 24-bit/96kHz, and 24-bit/19kHz depending on the source file.

Atmos playback through compatible TVs, Soundbars, and AVRs is easy to set as long as all your devices in the play-through path are compatible with Atmos and you use a high speed HDMI cables.

GETTING HI-RES MUSIC FROM STREAMING SERVICES

You can plug earbuds or headphones straight into your phone or computer, or use Bluetooth and listen to streamed music, but you can easily get better lossless hi-res audio playback by connecting your phone or computer to an external DAC via USB.

If you use an iPad for your streaming rather than a computer, or even a dedicated streaming device, the bitperfect transfer of music works flawlessly:

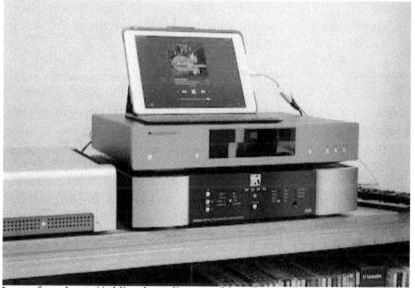

Image from https://addictedtoaudio.co.nz/blogs/how-to/how-to-get-the-best-out-of-apple-music-high-resolution-lossless

You will be able to add a Smart Keyboard Folio, for typing lookup requests, to the right iPad model (iPad Pro 12.9-inch (3rd generation or later), iPad Pro 11-inch (all generations) or iPad Air (4th generation or later). Then experiment with the different music streaming services, including Apple Music, and probably a container music player like Roon or Audirvana. You will have the easiest bug-free guarantee of bitperfect playback, whereby your player auto-adjusts to the sampling rate of the streamed music. Having said that, using container media players Apps like Roon,

Audirvana, and Amarra, explained below, manages bitperfect playback on other computers and Android devices. On their own streaming service apps do not do this (2024).

Yes, there are some Streaming DAC devices that include DACs in which case you do not need an external DAC (as you do with an iPad, iPhone, Android phone, tablet or computer). When you are concerned with convenience, that can be a good way to go.

Just using the headphone jack without an external DAC is limited. Most Android phones resample Audio output via their headphone jack at 16-bit/48kHz lossless/uncompressed, which is slightly better than CD quality, but not Hi-Res. iPhones are better with the headphone jack outputting 24-bit/48kHz lossless/uncompressed by default (the start of Hi-Res). PCs are variable depending on the internal DAC. But in any case the quality of the DAC in a phone or computer or TV will be noticeably subpar because they have a lot of electronics crammed into a small place, which creates noise on the audio signal. Using the headphone jack or Bluetooth is fine for Spotify and YouTube Music because their file format quality is about the same as those headphone jack outputs. If you are using a higher level streaming service like Tidal or Qobuz, you need an external DAC to gain the benefit of their higher resolutions.

Bluetooth also has its limitations. Bluetooth is slightly lower in quality than plugging headphones in because Bluetooth compresses audio so you end up with something a bit lower than CD quality. Newer Android phones with LDAC, compresses to 32-bit/96kHz at up to 990kbps. That sample rate sounds good, but because of the compression it will in reality be roughly at CD quality, which is 16-bit/44.1kHz; a little bit lower than what you get from the headphone socket itself. Again, Using Spotify and YouTube Music should work well enough via Bluetooth.

When you play music, it is best quality if done with **automatic sample rate matching/switching** to ensure the bitperfect integrity of the Lossless file. This means there is no unnecessary digital processing and noise added. The best way to do that is to use iOS, on an iPhone or iPad, and apart from Apple Music, use a container media players like Roon, Audirvana, or Amarra Luxe, which in-turn passes audio in a concentrated bit-perfect way, optionally upsampling for greater noise removal to improve the sound further no matter the Operating System. Streaming Service Apps typically do not sound as good as the above three **container media players Apps**. Here is a very general price range of those container media players, the OS they work on along, and what streaming services they integrate:

$$$ **Roon** (MacOS, Windows OS, Linux (x64), Roon OS) [Tidal, Qobuz, KKBOX] https://roon.app/en/.
$$ **Audirvana** (iOS, MacOS, Windows OS, Android OS) [Tidal, Qobuz, Hi Res Audio] https://audirvana.com/.
$ **Amarra Luxe** (MacOS and Windows OS) [Tidal, Qobuz,]
$ **Amarra Play** (iOS) [$ Apple Music, Tidal, Qobuz]; and acts as remote for Amarra Luxe) https://www.amarrasound.com/amarra-luxe/.

THE CONTAINER MEDIA PLAYER APPS will all access local music files, including DSD files, which you might store on your streaming device or an external drive.

I use Amarra Luxe on my HP ProBook laptop (Windows OS) because it is a reasonable once only payment, and I think sounds the best, even though its features are fewer. I use Tidal, but I open and close my account to save money, and buy music that I really love, which when stored locally I can access through Amarra Luxe or JRiver.

I also use a media player called JRiver, see https://jriver.com/. It is limited in terms of accessing music services to Amazon Music, but, rather than using it to access music services, it accesses free shared music, shared my members on Cloudplay, and I also have my local music file collection. I like its integrated DSP studio which allows me to set up EQ for my headphones using my preferred plugin (discussed in the EQ chapter).

I suggest trialing them all to compare the sound, features, and cost, and to trial different streaming services, to ensure the combination works properly on your setup. Just check the setup instructions such as the need to allow it to pass through any antivirus firewall you have on your phone or computer.

Here are the links to the streaming services, which you can incorporate into the container media players (if you have an account with them). Availability depends on your country:

https://tidal.com/. Have a listen to Tidal Masters, whereby the recording process was exceptional.
https://www.qobuz.com/. Qobuz has more Hi-Res than other services.
https://www.kkbox.com/intl/. Big in Asia.
https://music.apple.com/. Apple music is cheaper with iOS and MacOS devices.

So, where have we arrived at? Sure, if you are listening to streaming services like Spotify or YouTube music, the standard headphone jack or Bluetooth connections from a smartphone are suitable. For Lossless Hi-Res music you are far better off adding an external DAC. Even CD quality 16-bit/44.1kHz will sound a lot better in Lossless format on an external DAC. IOS devices provide the cheapest streaming service option of Apple Music. I also suggest trialing container media players. With container media players, you can easily stream bit-perfectly from a computer, or Android smart phone, accessing all the music services, except for Apple Music. For headphones, use open or semi-open designs where possible.

DACS

IN ORDER TO GET HI-RES OUT OF A PHONE OR PC COMPUTER, you commonly pass the audio via USB, HDMI, or optical cable to a Digital Analog Converter (DAC). DACs decode the bits, the bit depth, the sampling rate, any compression codec, and turns it into an analogue electrical signal (from numbers to a voltage varying electrical signal, matching the pressure of sound waves.

A DAC might be a dedicated DAC/AMP, in a TV, in a Soundbar, in an AVI, or in a Digital Interface. It is best to use a dedicated DAC rather than one built into another device.

MANY DAC/AMPS HAVE LINE LEVEL OUTPUTS. Line level outputs pass the analog signal, boosted to the Line Level and feed on to a Power Amp. Power Amps can be in Powered speakers, Soundbars, AVRs, or be dedicated Power Amps.

Not all DACs and AMPs are created equal. For example, a computer or TV DAC may introduce some distortion due to jitter (timing errors). To compensate for this, quality digital clocks are included in many external dedicated DACs or DAC/preAMPS to ensure the timing translation from digital to analog is error free. When you want to take advantage of higher quality audio, the DAC really matters, because the DAC conversion process is where the benefit of Hi-Res file formats is initially realized. Because of that it is always better to step around a TV, computer or phone DAC, in which case you use a digital cable (such as USB, Coaxial, or HDMI, or Optical) connection out to your external device that has its superior DAC ready to convert the digital to analog.

The quality of the **D**igital to **A**nalogue **C**onversion (DAC) process is critical for sound quality, but, so also is the after processing to your headphones and/or speakers, whether volume attenuation or true analogue volume control is used.

DACs and **volume attenuation** (not true volume control) is usually handled together, so you can adjust the volume to a Headphone jack on the DAC. By volume attenuation, I mean the volume adjustment is handled in the digital domain as apposed to adjusting the signal volume/strength in the analog domain. Volume attenuation has a downside in that it, by discarding bits, can add noise. True volume attenuation in the analog domain genuinely increases and decreases the signal strength, allowing for more control of quieter passages of music.

Many DACs are referred to as DAC Preamps, yet the headphone out is actually volume attenuated. When they are talking about pre-amplification they are talking about the signal they pass on to the next device, probably a Power AMP. One way to know if your DAC has true pre-amplification for its headphone jack is there would be an analog volume control. Look for information in the DAC's specifications or manual mentioning an "analog volume control" or "resistor ladder network." And terms like "stepped attenuation" or "analog pot". I bear this in mind when considering how DAC headphone amplification is handled. Most of the DACs I list below are quality products, but do not have true volume attenuation. This means to get the best out of your headphones you need to consider a dedicated Headphone Amp. We shall get to that. But for the time being we will focus on the DAC processing.

TYPICALLY A DAC/AMP will have file format specs that it can play up to.. Often this is up to 24-Bit 192 kHz stereo playback, and very often MQA (which is a compression unpacking method used mostly by Tidal streaming service), and up to Native DSD512. Those specs are really high, higher than you could ever want. Really, for most quality equipment lossless 24-bit/96kHz and native DSD64 is as good as it gets, and to gain more benefits you have to start spending more on all your gear. I have added a section on DSD and how it fits in at the end of this chapter. For more information of all file format specs, see the chapter 'Streaming and Play-through'.

A DAC, through its processing, has an effect on audio clarity and tone, and one big factor in this sound signature is the **DAC chip and associated circuitry**. I refer to these typologies below, and in the 'Streaming and Play-through' chapter. In brief they are:

- **Sigma-Delta**: most common, very accurate (some say too accurate), and cost effective.
- **R2R**: Uses resistors, connected up into a circuit, the technology being analogue and considered more musical sounding.
- **Multibit**: A digital equivalent of R2R, likewise considered musical sounding.
- **FPGA**: circuity designed for jitter-free/distortion-free speedy audio processing, resulting in a more analog musical sound signature, but costs more.

NOTE: **Tubes**, likely used after the DAC circuitry, also add a distinctively warm and classically musical sound signature.

Here are some **DAC** optons (with a focus on the chips):

IDEA 1 for an on the road Portable DAC/(which connects via USB to your phone's USB port):

ONE WAY TO GET HI-RES for portable use is to add a pencil DAC/(headphone preAMP), like the **Dragonfly Cobalt**, about $400, see www.audioquest.com/page/aq-dragonfly-series.html. It is plugged into a laptop, Android or iPhone, and the Cobalt version will maximally handle lossless 24-bit/96kHz, and MQA. The max limit will allow for the majority of Apple Music and Tidal content, but not so much Qobuz which has a higher percentage of even higher resolution content:

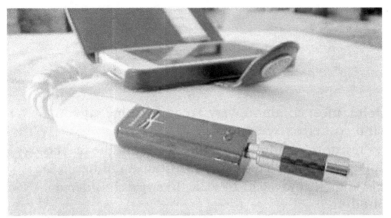

You might us an *iPhone or iPad with iOS 14.6 or higher*, running Apple Music (or Amarra Play integrating Apple Music) being an ideal choice for on the road. 2nd choice, would be an Android phone running Audirvana perhaps with Tidal or High Res Audio. You could also try Amarra Luxe with Tidal.*

Use the Apple Lightning-to-USB camera adapter for iPhones and iPads.

Use an Audioquest DragonTail USB adapter, which comes packaged with the Cobalt, to activate USB Host Mode on an Android phone, or another of the many OTG adapters.

* For iOS 14.6 / iPadOS 14.6, that will be iPhone 6s and later, but as above I also suggested to start from iPhone XS or later to get Atmos playback. For iPads, go for: iPad Pro 12.9-inch (3rd generation or later), iPad Pro 11-inch (all generations) or iPad Air (4th generation or later). Those iPads will also play Atmos (Apple

Spatial Audio) automatically, and you can use the Smart Keyboard Folio with them.

The Dragonfly Cobalt uses an ESS Sabre chip, which is something like a sigma-delta chip – clear and arguably the best kind of chip for electronic and punchy pop music. The Dragonfly Cobalt will show the playing sample rate by LED lights, namely: Green: 44.1kHz, Blue: 48kHz, Yellow: 88.2kHz, Light Blue: 96kHz, and Purple: MQA.

I RECOMMEND the **Qudelex T71 DAC AMP**, for about $370 [32-bit/864kHz, very good 20 band EQ], see https://www.qudelix.com/products/copy-of-qudelix-t71-usb-dac, which works for iPhone or Android:

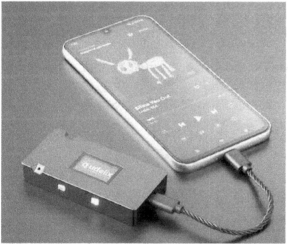

The iOS/Android mobile APP that goes with it allows you to enter the detailed EQ to suit your headphones – see the chapter on EQ for more details on where to get the EQ settings for your set of headphones. It is an important feature for a mobile DAC that will improve sound quality.

The **Chord Mojo 2** about $700, would be an upgrade from the above two mobile DACs, with its FPGA typology, heading more so in the direction of analog sound reproduction. [It will play up to 32-bit/768kHz, DSD256], see https://chordelectronics.co.uk/mojo-2:

To connect the Chord Mojo 2 to a Power Amp for desktop use, there is a nifty way to set the headphone out to line level. Press-hold both volume buttons when you turn it on. Do this each time you turn it on because it reverts to headphone amp when it is turned off. Use a cable that is 3.5mm male to 2 RCA, which connects from the headphone jack to a Power Amp. That makes it possible to use it as a mobile DAC and also as a desktop DAC passing audio to your Hi-Fi system.

* When you see sample rate specs higher than 192kHz; higher than hi-res sample rates from streaming services, the advantage is that a container media player like Audirvana can upsample interactively using your DAC. Upsampling, if done well, improves noise removal.

The **iFi Audio Micro iDSD Black Label Desktop DAC and Headphone Amplifier**, about $900, is another fairly portable DAC/AMP that can handle native DSD [Native DSD256, 24bit 352.8kHz (DXD384*) MQA], see https://ifi-audio.com/products/nano-idsd-bl/:

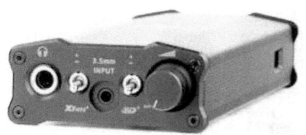

Similar to the Chord Mojo 2, there is a Line out facility for on-connecting to a Power Amp. In that case use a cable that is female RCA to RCA male. The headphone volume is by digital attenuation, just to be clear.

* DXD384 is the same as 24bit 352.8kHz, which is often used for recording and/or post production processing, before mastering to the DSD64, 128, 256 516 formats.

IDEA 2 For convenience, use a Streamer DAC/headphone DAC/(pre)AMP:

FOR CONVENIENCE, WE HAVE A NUMBER OF INTEGRATED STEAMER DACS. Possible downsides are that Apple Music is not integrated into them, and you will be paying more for a Streamer DAC. On the positive side, because they are integrated they should work without you running into buggy problems, and there is the convenience. Here are some streaming DAC options:

- **FiiO R7 All-In-One Streaming Amplifier**, about $800, see, https://www.fiio.com/r7 [Roon ready, integrates any Android streaming App, DSD64/128/256 converted into PCM, remote controls]

For remote control, you can use the FiiO Music App for Android or iOS. That is a pretty good system for a desktop setup. There are optical and coaxial digital outs in case you want to hook it up with an external Power amp or AVR.

- **Cambridge Audio CXN V2 Network Streamer DAC**, about $1,200, see https://www.cambridgeaudio.com/usa/en/products/hi-fi/cx-series-2/cxn [Roon ready, Spotify, Tidal or Qobuz, DSD64 converted into PCM, remote control].

For remote control, use the The StreamMagic app for Android or iPhone. Connect on to a Power Amp. It can upscale music to 24-bit/383kHz, being a way to remove noise during the DAC processing.

The Cambridge Audio CXN V2 Network Streamer DAC does not have a headphone jack, but I would argue that that is a good thing. Then you might be utilizing a dedicated Headphone amp with true volume pre-amplification, or an Integrated amp with a headphone jack (though the integrated amp probably will not have true volume pre-amplification).

- Matrix Audio Mini-i Pro 4 Music Streamer, DAC & Headphone Amplifier, about $1,400, see https://www.matrix-digi.com/product/94/mini-i_Pro_4. [Tidal, Qobuz, High Res Audio, remote control | DSD2.8MHz, 5.6MHz, 11.2MHz, DSD22.4 Native – transferred by DoP].

Use the MA Remote App. Connect on to a Power Amp.

The Matrix Audio Mini-i Pro 4 Music Streamer, DAC & Headphone Amplifier says it is a headphone amplifier, so let's see if it uses true analogue volume pre-amplification for the headphone jack. The website says it has a fully balanced amplifier circuit design. The benefit of balance headphone jacks are more for long headphone cables, otherwise the improvement is debatable. There is nothing in the manual that describes true analog attenuation, so we can understand that it uses digital volume attenuation, thus headphone playback can be improved otherwise.

- Denon AVR that has in-built HEOS, preferably with Class AB amplification:

For remote control of Denons with HEOS, use the HEOS App.

HEOS makes it possible to nest purchased streaming services like Tidal.

AVRs mostly use digital volume attenuation. Typically, if the volume control has only 270 deg rotation - that's analogue. If it turns round and round with a display that shows a number then it's digital. In my case I presently have a Denon AVR X220W, and that has the spinning digital volume control.

Note 1: Modern A/V receivers usually do better with higher impedance headphones, like 250-Ohm to 600-Ohm. Indeed my Sennheiser HD600 has 300-Ohm impedance, and does sound very good out of my Denon AVR X2200W, which uses the Burr Brown PCM1690 8 channel Delta-Sigma DAC chip, and class AB amps. Newer AVRs will handle Hi-Res audio, Surround Sound and Atmos, and DSD. [At the moment (2024) my Denon AVR X2200W plays up to 24-bit/192kHz [and DSD64 through a plugged in flash USB].

Note 2: It may be you want to stream music from your AVR through a separate Power Amp to your speakers for the advantage of better amplification in the separate Power Amp, in which case you need an AVR with Pre-outs. That level of AVR will also have good in-build Power Amps so to gain advantage you need a pretty good separate Power Amp.

Note 3: If you are passing music through an AVR, you will have a number of mode options. A common one is Direct mode, another is HT Bypass (on higher end models). These isolate surround sound, ignoring the home theater processing, and other features that are not needed for stereo play-through, thus the sound will be cleaner.

IF YOU ARE PLAYING FROM A NEWER SMART TV, you will commonly stream to your AVR via an HDMI ARC or eARC connection (with a high speed HDMI cable), thus getting Hi-Res and any Atmos (if your AVR or Soundbar is Atmos compatible). That will probably be with HDMI 2.1b which supports 24-bit/192kHz (though movies will usually stream at 24-bit/48kHz).

- The Musical Paradise MP-DX ES9068 ES9039 Hi-Res Music Streamer Tube DAC, for $969, *[DAC with Tubes in the output stage]*, see https://www.musicalparadise.ca/store/index.php?route=product/product&path=61&product_id=109. It comes with built-in streaming using Rasberry Pi and Volumio. I inquired about this setup to the developer, Garry Huang. He said Volumio does not sound the best, and instead of using Volumio, suggests configuring Ropieee for Raspberry Pi on the MP-DX, then with a PC to run Audirvanna or hqplayer, both of which now recognize Ropieee.

So, if you are happy to tweak that system, the MP-DX Tube DAC will provide that extra Tube Sound, and as with all Music Paradise products, you can upgrade internal parts without needing to solder. That seems a good idea, to me, if you are going to on-connect a Topping type of Headphone Amp (as introduced below) and connect on to an Integrated Power Amp (and are happy to set up a Ropiee setup). I do explore below, in more detail, the Musical Paradise Pre-amplifier, which I think is a better way to access tube amplification typology.

A word about (vacuum) Tubes.

The story on Tubes goes something like this, starting with a quote from Charles Whitener, the CEO of Western Electric (1996-2008), the tube-and-electronics manufacturer: "Music played through transistor electronics has a certain compression and thinness, while a vacuum-tube system has a voluptuous, musical soundstage with front-to-back depth. It's lush, it's real... it sounds human."

Now, things have moved on since then. Modern improved transistors are commonly used in audio gear like DACs, Preamps and Power AMPs. Tubes amplification has also modernized, so while the classic Tube sound is quite colored, modern tube amplification tends to be more neutral sounding.

Tube rolling (changing tubes) is also a thing, where different tubes are swapped in to replace off the shelf ones, each tube having a slightly different sound signature. They vary from neutral and clean sounding to more warm sounding, as do the basic designed sound signature of the Tube Amps. Thus you might choose a tube amp according to your listening preference (classical or modern sounding) and according to your preferences you might roll in different tubes. The second part of the original quote is still true today, "(tubes have a) voluptuous, musical soundstage with front-to-back depth. It's lush, it's real... it sounds human."

While there are a few DACs (like the Musical Paradise ones above) that use some form of Tube technology integrated with DAC chips and/or transistors, tubes are more often implemented in dedicated Pre-amplifiers and Integrated Power Amps. While some dedicated preamps and integrated amps are solely solid-state components (meaning designed with a combination of transistors, tiny-chip integrated circuits, diodes – any non moving component), others combine those technologies with Tubes for a more modern tube sound that is more neutral and clear compared to the greater bloom characteristic of classic tube amps.

The tube difference is a kind of holographic euphonic, that is very pleasing for vocals and acoustic instruments, and more relaxing for longer periods of listening. Modern Tube applications tend to be saturated and fast, so you really are getting close to an analogue being there experience. Then, depending on tube selection, you can shape that sound more in the direction of classic tube sound or neutral, depending on your preference.

A word about clean and neutral audio. This is the holy grail of audio design – to have equipment that gets out of the way, so you hear music, with all its life, no-matter the type, as it sounded in the studio.

BEFORE SUGGESTING more common approach of DAC/(headphone)AMPs, let us consider how a Digital Interface can also be used as one. While a Digital interface (DI) has an Analog to Digital Converter (ADC) for digitizing recorded audio, a DAC and headphone preAMP are in there too, so they can be used, as a DAC/(headphone preAMP).

A digital interface, the **BabyFace Pro FS**, about $1,000, see www.rme-audio.de/babyface-pro-fs.html uses Delta-Sigma type of chip in their own circuitry variant, which will play up to 24-Bit 192 kHz [no DSD or MQA]. The BabyFace Pro FS is fast and neutral, being what you want in a good DAC.

There are other recommended Digital interfaces in the previous chapter, all of which can be used as DACs.

An advantage is it may well have EQ processing built-in. I find that particularly useful for applying EQ when listening to headphones, and it is easy to get the EQ settings online for any particular model of headphones. See the chapter on EQ for more details.

For a listening only preference, the RME alternative, with a few more bells and whistles, is the RME ADI-2 DAC FS, see below.

The limitation of the RME BabyFace Pro FS is that unlike the ADI-2 DAC FS, it won't have MQA and DSD playability. I do not think this will be an issue for most people, because the Tidal streaming service is moving away from MQA and not many people are pursuing DSD, while at the same time lossless hi-res at 96kHz or 192kHz sampling rate is becoming more common across most music streaming services. For DSD there can also be other playback options, such as a SACD player, and many AVRs accept DSD on flash USB drives.

IDEA 4 Use a DAC with a Delta-Sigma chip:

DELTA-SIGMA (sometimes called Sigma-Delta) is the most common, and is very accurate sounding, some believe a little abrupt or complain about 'digital glare'. It is well suited to modern punchy music.

Delta-Sigma is the most common and cost effective DAC chip type. The D/A process includes over-sampling into higher frequencies to reduce the noise created by the conversion (which is how you benefit from higher spec formats like 24-bit/96kHz, as explained in the chapter 'Clean Audio'.

Delta-Sigma chips are made by ESS, AKM Semiconductor, Analog Devices, Texas Instruments, and Wolfson

Microelectronics.

Here are some quality options that utilize the Delta-Sigma chip:

- Topping DX5, about $450, see shenzhenaudio.com/products/topping-dx5-dac-mqa-2xes9068as-decoder-dsd512-pcm-32bit768khz-high-performance-audio . It plays up to 32-bit/768kHz [and DSD512, and MQA].

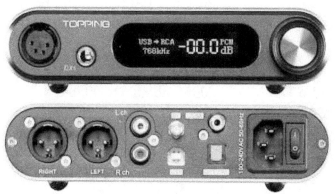

Connected via USB from your Windows, Android, MacOS, iOS device.

- **Fiio K9 Pro ESS**, about $850, see fiio.com/k9proess . It plays up to 32-bit/384kHz [and DSD256, and MQA].

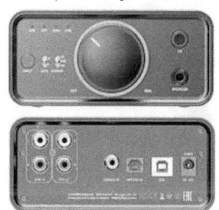

Connect via USB from your Windows, Android, MacOS, or iOS

device.

- *Schist's Jotunheim 2 with added DAC module* (or select the Passive Phonon module), about $500, see schiit.com/products/jotunheim-1. It plays up to 32-bit/192kHz [no DSD or MQA]. The lack of DSD or MQA support need not put you off. You only need MQA if you are gong to use Tidal as a music payer (and, as of late 2023, TIDAL is moving away from MQA in favor of uncompressed lossless FLAC files), and bear in mind you will only be interested in DSD if you are seriously into acoustic, jazz and classical music, because most of the DSD material is in those genres.

- **Cambridge Audio DacMagic 200M**, about $1,000, see cambridgeaudio.com/row/en/products/hi-fi/dacmagic/dacmagic-200m. [also MQA, and up to Native DSD512].

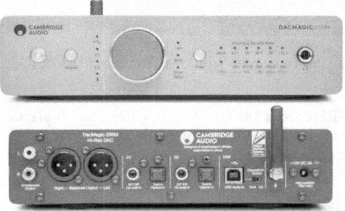

Connect via USB from your Windows, Android, MacOS, or iOS device.

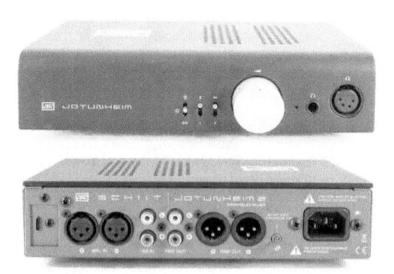

Connect via USB from your Windows, Android, MacOS, or iOS device.

- **RME ADI-2 DAC FS,** about $1,300, srme-audio.de/adi-2-dac.htmlee . Compared to the RME BabyFace Pro FS above, it has fewer inputs than a digital interface, above, it adds listening focused features (increasing from 3 to 5 EQ bands, Crossfeed adjustment for nicer headphone listening, and comes with a remote control). And, it plays up to 32-bit/768kHz [and DSD256, and MQA].

Connect via USB from your Windows, Android, MacOS, or iOS device.

IDEA 5 Use a DAC with a R2R resistors:

R2R/MULTIBIT IS MORE MUSICAL SOUNDING THAN DELTA-SIGMA. R2R uses a series of resistors (resistor-ladder) for the D-A conversion, with no pre-conversion to PCM. Multibit parallels the same process, but digitally. R2R/Multibit DAC manufacturers include Aqua Acoustic Quality, Denafrips, Holo Audio, Metrum Acoustic, MSB, Schiit, and Totaldac.

The **Denafrips Enyo** (formally known as the Ares II), for about $850 is a high quality and affordable **R2R**), denafrips.com/enyo. It plays up to 24Bit/1536KHz [and up to DSD1024, no MQA]:

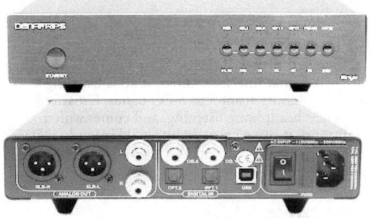

Connect via USB from your Windows, Android, MacOS, or iOS device.

See this review, by A British Audiophile, see youtu.be/bb-O8JF24PI. Since it does not have a headphone AMP, you would need to connect on to one, or to an integrated Power AMP which has a headphone jack and volume control, or an AVR. A dedicated headphone amp/preAMP. A dedicated Headphone amp will have true analogue volume control for best low volume listening.

Here is a Multibit DAC (relatively digital equivalent of R2R suggestion:

- The **Schiit Audio Jotunheim,** about $600, uses Multibit technology. Multibit is a somewhat digital version of R2R, also providing a more analog type sound, see https://www.schiit.com/products/bifrost. It plays up to 24-bit/192kHz [no DSD or MQA]. The lack of DSD or MQA support need not put you off. You only need MQA if you are gong to use Tidal as a music payer (and, as of late 2023, TIDAL is moving away from MQA in favor of uncompressed lossless FLAC files), and bear in mind you will only be interested in DSD if you are seriously into acoustic, jazz and classical music, because most of the DSD material is in those genres, or, you have had a SACD player and have digitized or are planning to digitize your SACD disks to DSD files.

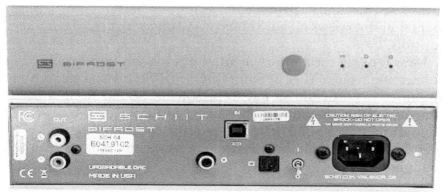

Connect via USB from your Windows, Android, MacOS, or iOS device.

You might have noticed, the Topping DX5 and Schiit Audio Jotunheim include a 4-pin XLR balanced Headphone jack. Similar to XLR interconnect cables but 4-pin, being balanced, the idea is to remove any noise introduced to the actual cable by doubling the electrical signal at the start of the cable, and canceling any difference between the two signals at the end of the cable. This sounds like it will be an improvement, but actually it is only an issue to resolve if headphone cables are more than 10 feet long or your environment is plagued by heavy electrical gear like a professional audio production environment.

HEADPHONE AMPS

A DEDICATED HEADPHONE AMP WITH TRUE ANALOGUE VOLUME CONTROL WILL IMPROVE HEADPHONE LISTENING. It is not that digital volume control is inferior in quality DACS, but that a combination of digital volume attenuation in the DAC and analog volume control both have their advantages, so using both is better. For a technical understanding of this see this article by S. Andrea Sundaram at Soundstage Ultra, https://www.soundstageultra.com/index.php/features-menu/gener al-interest-interviews-menu/311-what-s-wrong-with-digital-volume-controls.

I was looking for a mostly neutral headphone amp for on-connecting from a Tube Pre-amplifier (MP-701). The Tube pre-amplifier was providing a way to personalize sound along with benefiting from the holographic aspect of tube sound. I was on-connecting it to a Solid State Power Amp, to neutrally amplify that sound for the speakers. I was looking to do the same with a headphone amp – to do so neutrally, perhaps add a little dynamic.

I looked at the **Topping A30 Pro**, for about $250. It is neutral, adding a little dynamic, with gain switching for better volume control, also includes 4 pin XL balanced connection if desired, no pre-amp function), generally for high impedance headphones, like my Sennheiser HD600s, see this review by GadgetryTech, https://youtu.be/oG7ySCpqD4o

If you like, you can connect it between a DAC and a Power Amp, and when you tun it off the signal passes through it.

For an even better result, you could opt for the **Topping A70 Pro**, for about $500, with it utilizing R2R in a hybrid way with its Digital Stepped Attenuator, thus realizing accurate sound at low volumes. It also has a trigger, so I could trigger it to turn on from a high spec AVR (one less button to worry about):

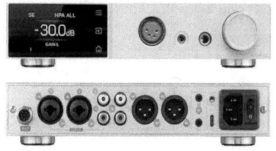

Stepping up in sound quality, try the **Aune S17 Pro**, for about $700, neutral like the Topping, but a more neutral, dynamic and ric engaging sound, especially when warmed up. And it adds fully balanced technology, so here if you want to use balanced headphones, it is worth it. See this review by Soundnews, https://youtu.be/wiISyB-HcxQ. For a review which details the technological reasons for the improved sound (twin JFET, Class A, adjustable current, R2R volume control) see this review by iiWi Reviews, https://youtu.be/r6Ku5PhYXME

The next step up, probably is the **Schiit Mjolnir**, about $1,200 (so if you are on a budget that would have to be a second hand pickup. see https://youtu.be/izwxsRnLNec. Here is the manufactures shop with all the details, https://www.aune-store.com/en/aune-s17-pro_110409_1235/

The **Schiit Lyr+** is a Tube Headphone Amp, for about $650, or $600 without a tube, see https://www.schiit.com/products/lyr:

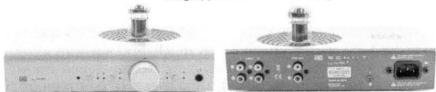

And, by the way, the volume control is true analog. It also has the option of switching between Solid State and Tube modes, so you can easily compare the difference. Here is a review from Passion for Sound, https://youtu.be/mDVH7QW_RGY, recommending the **Elecro Harmonics 67N7 Goldpin**:

PRE-AMPLIFIERS

IN THIS SECTION I AM GOING TO DISCUSS TUBE PRE-AMPLIFIERS, not solid state pre-amplifiers, because it is an ideal place to introduce tube sound into your in-room hi-fi stereo playback path.

A dedicated Pre-amplifier often uses analog volume control, which can be an advantage combined with the Digital or R2R volume control of your DAC. The tube part of it gains the benefit of the holographic nature of tubes, sound matching to your speakers, and to your personal preference.

A preamplifier goes before a Power Amplifier, or is inside an Integrated Power Amp (which is why it is called integrated). Here is a discussion by Paul McGowan of PS Audio explaining that quality dedicated preamps improve music playback quality, see https://youtu.be/Dh27E7YKN9s.

There are a few **variations to pre-amplification.** If you record audio through a passive ribbon microphone, because the signal level is so weak, it is common to feed the microphone into a ribbon pre-amplifier to do a specialized job of increasing the level and managing the impedance (which affects tone). Likewise, a phono player has a weak initial signal, and so needs a specialized phono preamp to increase the level and ensure balance and accuracy. That may be inside the phono player or may be in an Integrated Amp, or in a dedicated Phono Pre-Amp.

AVRs have multiple internal preamps connecting on to line level Power amps for each speaker output. Higher level AVRs have matching RCA preamp outputs for each speaker output, as an option to ignore the AVR's Poweramps, popularly used for on-connecting to a better quality Power amplifier.

If you connect up a dedicated pre-amplifier between your DAC and Power Amp, the question of **how you manage volume**

control comes up. What you want is your DAC to send the best quality audio to your preamplifier, and that will be at about 0dB, probably about 2pm on the volume control, which is handing you the audio in its original form.

You could experiment – basically you open up / turn to 2-3pm on your best volume control (which mean you open up your best pre-amplifier), and use the volume controller on your other device. Paul McGavin of PS audio suggests you turn the volume to max on the pre-amp (0 dB) and use the DAC to control the volume, see https://youtu.be/HdytsbrzjCA. This is based on the Pre-amplifier being better than the DACs digital volume attenuation, so by maxing out the dedicated pre-amp volume you are opening it up to fully utilize its pre-amplification circuitry. If, however, you do not have a dedicated pre-amplifier between your DAC and Integrated Power Amp, then set your DAC to max (0 dB) and control volume on your Power AMP /Integrated AMP.

By way of introduction, the **Schiit Freya+** and **Musical Paradise MP-701 tube preamp** are two examples of reasonable budget dedicated tube pre-amplifiers.

Schiit Freya+ dedicated pre-amp   MP-701 pre-amp
True analog volume control? Yes.   Yes.

Schiit Freya+, see https://www.schiit.com/products/freya-n
Musical Paradise MP-701, see
https://www.musicalparadise.ca/store/index.php?route=product/product&product_id=103

There would be a problem if you tried connecting a dedicated preamp like the Schiit Freya+ or Musical Paradise MP-701 mk2 into an integrated amp or AVR, because Integrated Amps and AVRs have their own internal pre-amps and you do not want the unnecessary processing, and so added noise, of a preamp into a preamp. Therefore, a dedicated Pre-amplifier needs to be connected to a dedicated Power Amp.

NOTE: You mostly cannot step around the preamps of integrated amps or AVRs, even if you have HT-Bypass. HT-Bypass in an AVR steps around the home theater processing but not around the Preamp or Poweramp. A Pure Direct setting on an AVR also does not step around the AVRs Preamp or Poweramp.

The **Schiit Freya+** dedicated pre-amplifier, for about $1,000, or $890 without any tubes on initial purchase, is a pre-amplifier that will introduce modern tube technology which is fast, clean, comparatively neutral sounding, and holographic.

The Schiit Freya+ also has a passive mode, that steps around the tube path for a comparatively more clinically accurate solid-state flavor. You might choose to use the solid state mode for more punchy electronic music.

The Schiit Freya+ also accepts other inputs such as a phono or CD player, see https://www.schiit.com/products/freya-n. See this review by Darko Audio, https://youtu.be/5iAmj_ptVaw. And, in this next review by Thomas & Stereo he suggests rolling/swapping in not too expensive, matched quad GE 6SN7 GTA or GTB tubes, at about $150, see https://youtu.be/LwMXMtSprxI. The GE 6SN7 GTA are for clear yet smooth sound; extended highs and an open and airy top end. The **GE 6SN7 GTB**s are more neutral with a faster and tighter bass response:

Thomas & Stereo also reviews the **Musical Paradise MP-701 tube preamp,** for about $1,200, or $720 without any of the tube options on initial purchase, see https://www.musicalparadise.ca/store/index.php?route=product/product&product_id=103.

Thomas of Thomas & Stereo judges it as similar to the Schiit Freya+, off the shelf, but that with upgrades it can achieve a Hi-End sound. It is designed for parts swapping, without soldering, just loosen and screw down, with a manual to make sure you flip the right switches.

Output tube

Coupling Capacitors

Output tubes

Power filtering bypass capacitor

Rectifier tube

This preamp responds well to tube rolling, which is an indication of fairly good design. As per Thomas & Stereo suggestions, see https://www.youtube.com/watch?v=xbvxtiOeJv4, or I Am Mad's suggestions, see https://youtu.be/XxqtiDKJA0E. https://youtu.be/XxqtiDKJA0E.

Here is an owners discussion forum, see https://www.stereonet.com/forums/topic/312729-musical-paradise-mp-701-mark-ii-owners-discussion-thread/

Here are some suggested upgrades for the MP-701:

(a) change the **Output tubes** to x3 **Genalex Gold Lion E88CC /**

6922 preamp tubes [for focused and solid mid-range, expanded soundstage, and air presence; a neutral transparent sound], for about $220:

Another option recommended by I Am Mad for 2-way and 3-way speakers, is the **E88CC-Performance-6922WA-S4A** (a Dutch version of the E88C) [smooth, punchy, detailed and clear], for about $160 for three:

(b) change the **Rectifier tube** to x1 **Genalex Gold Lion U77 / GZ34 rectifier tube** [singer detached from instruments, and hi-end sound stage performance], for about $60:

OR, x1 **Mullard CV1377 KB/QDA GZ34 5AR4** rectifier tube [for more excitement - faster attack, shorter decay], for about $200.

Comparatively, I Am Mad, as per video link above, recommended the 5T4 steel [earthy, and forceful] as an excellent budget option for about $30, or for more musical excitement, the **Linlai 274B Black Plate** [smooth dynamic, vibrant mids, high shimmer], for about $50.

Before slotting in tubes, clean the pins with DeoxIT Contact Cleaner and Rejuvenator to ensure the best contact: Just joggle the pins in and out of the socket a few times - that's it.

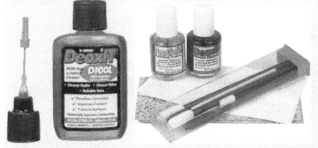

If you have gold pins or contacts, use the ProGold as seen here with a bottle of Deoxit. These are powerful cleaners and will leave a thin film that can help improve conductivity and prevent future corrosion.

(c) For the **Power filtering bypass film capacitor,** the owner Garry Huang recommends upgrading the stock Stock Bennic power filtering bypassing capacitor 350V 6.8UF. Capacitance can be 0.1UF – 10UF. Voltage needs to be >=350V. Nonpolar. to **Mundorf Capacitor 6.8uF 600Vdc MCap® Supreme Series Metalized Poplypropylene Axial,** see https://www.hificollective.co.uk/catalog/sup8180-68uf-600v-supreme-polypropylene-capacitor-p-1159.html

(d) FFor the Coupling Capacitors you will be pretty well off with the stock Obbligato gold coupling capacitor, which is 630V 2.2UF (Capacitance can be 1.5UF – 4.7UF. Voltage needs to be >=350V. Nonpolar). You could also upgrade on purchase to x2 VCAP ODAM 400V 2.2 UF for an extra $200:

Other options commonly discussed are the Mundorf Mcap Supreme EVO Aluminum Oil, for about $70, OR better than that, the Mundorf Mcap Supreme EVO Silver Gold Oil, for about $180.

Note: The higher the coupling caps uF value, the more emphasis on the low frequencies. If you want more bass, you put a higher uF value within the range for the coupling caps, if you want less bass you put a smaller uF value. Because of this you might benefit from upgrading your tubes and Power filtering bypass film capacitor first, then because each system is different, decide whether you want to bump the bass up or down. To that extent try a uF rating up or down from the stock uF2.2.

When working with capacitors, especially electrolytic capacitors. you need to discharge them before handling them. Here is a video on general safely when working on tube amps, by Premier Guitar, which also includes discharging capacitors, see https://youtu.be/DkEc58-vWc4. He uses a 100 Ohm resister held or soldered to one crocodile clip, being one one end of a wire with crocodile clips on each end. The end without the resister is clipped to the chassis. The he verifies the discharge with a voltmeter (set on DC volts) – negative lead to the chassis; positive lead to the positive end of the capacitor.

For a system that discharges and verifies in one go, I bought a YEKMLCO Digital Capacitor Discharger, which I bought through AliExpress.

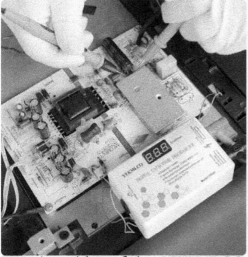

Here is a video of the YEKMLCO Digital Capacitor Discharger being used, https://www.youtube.com/watch?v=lNA5roVtU38

What about DIY Tube Pre-amplifier kits?

The best DIY kit I can find of a dedicated Tube pre-amplifiers (tubes slot in the back), that uses good parts and has excellent reviews behind it is the **Elekit TU-8500 Stereo Tube Preamplifier Kit**. Designed and parts from Japan, it has three Line-ins, a Phono in, and a Pre-out. For about $500.

Available at Tube Depot and check out eBay, see https://tubedepot.com/products/elekit-tu-8500-stereo-tube-preamplifier-kit. Here is an excellent six part review at cheaptubeaudio.blogspot.com, see https://cheaptubeaudio.blogspot.com/2015/03/review-elekit-tu-8500-full-function.html. Here is the manufacturer's site, see https://www.elekit.co.jp/en/product/TU-8500.

Here at Hi Fi Collective, you can also buy upgrade parts, and watch a DIY series of construction videos, see https://www.hificollective.co.uk/kits/elekit-audio-kits/tu-8500.html. For tubes, perhaps upgrade to x2 GE 6189 tubes for a little more air, more fine details and in general a more sophisticated presentation , or x2 clear-top GE 6201 (white letter) for less refined but better controlled bass.

I would recommend getting one from HiFi Collective, see https://www.hificollective.co.uk/kits/elekit-audio-kits-tu-8200r.html, and opting for the TU-8200VK version, being the one with the Lundahll LL2777B output transformers, for about $1200.

This preamplifier also pairs well with the **Elekit TU-8850** Pentode

Single-Ended Power Amp Amplifier Kit, discussed in the next section.

POWER AMPLIFIERS

POWER AMPS ARE OF COURSE INSIDE INTEGRATED AMPS, being a combination of pre-amplifiers and Power Amps. We have looked at AVRs that are also hybrid Integrated Amps focused on Surround Sound. Let us return to **dedicated stereo Power Amplifiers** that you connect to a DAC, Pre-amplifier, or Pre-out from an AVR. At the end of this section I will come back to Integrated Power Amps.

IN GENERAL, I SUGGEST THE IDEA OF USING A SOLID STATE POWER AMP RATHER THAN A TUBE POWER AMP. Firstly, I am following the idea of combining the different typologies, endeavoring to get the best out of all combined (so tube sound; that is pre-amplified up with neutral dynamic solid state). Secondly, a solid state amp can connect to a Two-way Power Amp to Speaker switch for integration with an AVR system (discussed below), whereas a Tube amp is not suitable for that because they need the speakers to be always connected when on to displace the power load that passes through the unit. But, understanding my approach, you of course have your own preferences, such as using an Integrated Amp with 'Main-in' connections and 'Direct' or equivalent switch (explained in the introduction to this section), you may not want to go down the path of adding a dedicated pre-amplifier, indeed adding a dedicated tube amp after an AVR (with pre-amp out connection) is the other common way to go, but starts to add up cost wise.

Firstly, here is a dedicated Tube based Power Amp. Because of the high cost of Tube Power Amps, here I mentioned the quality Japanese design and parts **Elekit TU-8850 Pentode Single-**

Ended Power Amp Amplifier Kit, minimally about $1200 (not including Tubes, which would be another $400).

Here is the manufacturers site, listed as PENTODE SINGLE-ENDED POWER AMP KIT [TU-8850], see, https://www.elekit.co.jp/en/product/TU-8850.

I would recommend getting one from HiFi Collective, see https://www.hificollective.co.uk/kits/elekit-audio-kits-tu-8850e.html, and opting for the **TU-8850VK** version, being the one with the Lundahll output transformers, for about $1700 + $400 for tubes. On that webpage there are also upgrade options for about $280. Plus freight costs, which might be as much as $200.

Because of the other **alternative** whereby you use an Integrated Power Amp that can integrate with an AVR, on the basis of having a 'Main-in' connections and 'Direct' or equivalent switch (explained in the introduction to this section), Here is an example of a Tube based Integrated Amp with those features, the **Musical Pardise MP-501 Tube Integrated Amp,** but its costly at about $1,500:

It has the Main-in connections in the back, and on top the "Direct" switch, in this case named, 'MAIN IN':

See the website at
https://www.musicalparadise.ca/store/index.php?route=product/
product&product_id=55

Now to Solid State dedicted Amplifiers

One approach is to **DIY**. I built a DIY Power Amplifier from Valutronic, a Swedish company, from whom I also bought plans and resources to build my ortho-acoustic speakers. The result was true to their claims of an audiophile level hybrid class AB PowerAMP, see https://www.valutronic.com/E/amp.html. It cost me a total of $700 to build the **PowerAmp 5 kit,** though I could have done it for $500. I paid $400 for the kit, $70 for a toroid transformer suitable for New Zealand 230V power rating, as advised by Valutronic. $100 for upgraded top quality internal wire from https://www.hificollective.co.uk and pure copper binding posts, and I optionally paid about $100 to get some acrylic sheeting cut and drilled as per the plans (rather than doing it myself), plus lets allocate $30 to soldering equipment:

Here are some dedicated Class A/B poweramps you can buy off the shelf:

- Clone FM300A Hi-End 300W HiFi Stereo 2.0 Channel Home Audio Power Amplifier, for about $400. It is a clone of the original Swiss made FM300A. Here is a review by I am Mad, see https://youtu.be/KL346EfvKDw
You can access these by searching AliExpress:

- Schiit Vidar 2, for $800, see https://www.schiit.com/products/vidar2:

- **Audio by Van Alstine Vision Set 120 power amplifier,** for $900, see https://avahifi.com/products/vision-set-120-power-amplifier:

Integrated Power Amplifiers

NOW, TO CIRCLE BACK TO THE IDEA OF **SOLID STATE INTEGRATED POWER AMPLIFIERS.** You can listen through headphones from the Integrated Amp (although not being as good as dedicated Headphone Amps because it will be digital volume attenuation instead of true analogue, thus audio quality suffering slightly at low volumes). Have a look at the **Cambridge CXA81 Integrated Stereo Amplifier,** for about $1,000, see https://www.cambridgeaudio.com/usa/en/products/hi-fi/cx-series-2/cxa81

Why not an AVR?

INSTEAD OF CONNECTING TO AN INTEGRATED AMP you could connect on to an AVR (which is really an integrated amp with additional features for surround sound and video allocated pass-through). I presently (2024) have a Denon AVR X200W, which I bought second hand for about $200, and which handles Atmos and has Class AB amplifiers (for better music rendition). Lets look at a higher spec AVR that has better quality internal parts and has additional pre-outs (giving you the choice to add on separate Power Amps), and separate multi-sub outs (allowing for a best approach to multi-subs (discussed elsewhere in this book). I point to, as an example, from Onkyo's high end line, the **Intregra DRX-9.4**, for about $2,200 new, see https://integrahometheater.com/product/drx-5-4-9-2-channel-network-a-v-receiver/# . My approach is to wait until they come up second hand. As of 2024, second hand ones are starting to come up on on eBay. Another option to look for second hand is the Denon AVR-X4300H AV Amplifier.

While I have a Smart TV connected to my AVR I also connect my DAC via an optical cable to my AVR. In that case the AVR automatically recognizes an incoming stereo signal, sending it to the main Left and Right Channels, though you can select multi-channel stereo and other options, as you prefer, to utilize your

surround sound speakers capabilities. For a TV connected audio signal it will recognize any Surround Sound or Atmos format and distribute it to your multiple speakers accordingly.
at and distribute it to your multiple speakers accordingly.

Sometimes if you buy second hand from eBay, especially AVRs, you might be looking at products which are specifically power rated for another country, quite often 110V for America or 100V or 110V for Japan. Those countries have a large hi-fi market and so second hand availability is more common than elsewhere. For example, If you wanted to buy an Intregra DRX-9.4 from the US or Japan, and you live in Europe, England or Australia, you will have a power voltage issue. Check the power rating label, like this on the back of the device:

In this case it is a US rating. When the label is like this and you are in a country where the power rating is higher, such as Europe,

UK and Australia, you can add a **step down transformer** to step down from your wall socket voltage to the equipment voltage. The transformer should have max watts at least 50% higher than your appliance wattage. In the Integra's case, checking the specs, it is 850W for the North American model, so you would want a 1500W step down transformer, the bigger the better. You should be able to find one locally, perhaps for $100. That is a little extra money and an extra switch, so buying second hand from your own country or countries with your power rating is always preferable.

On the other hand, some second hand devices will have a built-in voltage selector, like this:

Many devices, though usually not AVRs, automatically adjust to the countries voltage, being something to check before purchasing internationally.

 ## HI-FI AND TV-AVR SYSTEM INTEGRATION

EACH TYPOLOGY HAS PROS AND CONS, so my approach is to think about utilizing a mixture of the typologies in your playback pathway. Above, we looked at the Denafrips Enyo R2R DAC, and adding on a Topping dedicated Headphone amplifiers that used R2R in its volume control or a Tube based Headphone Amp from Musical Paradise; we looked at a couple of dedicated Pre-amplifiers, one from Schiit Audio, and one from Musical Paradise, including DIY options; we also looked at some DIY and off the shelf dedicated Power Amp, also some Integrated Power Amps and an AVR. Let us now look at system integration whereby our Hi-fi system is integrated with a TV AVR Surround Sound system.

By system integration I mean integrating Hi-Fi system devices, and integrating that with a Smart TV/movie App to an AVR Surround Sound/Atmos system. Of course these two systems can be separate and would be if you have a separate listening room. I favor a family room approach, putting both systems in a lounge living environment.

THE MAIN WAYS I tie both Hi-Fi and AVR systems together is to (a) use the Smart 4K TV as an external computer monitor. In this way any Music Player, YouTube, online shopping or other work you are doing can be spread out on one big screen. When set correctly, a text on a 4K TV is just as fine as on a computer screen. The other tie-in-point is to share the main Left and Right speakers with the AVR system and the Hi-Fi system.

ANOTHER ASPECT OF HI-FI AND TV-AVR SYSTEM INTEGRATION is utilizing the benefits of the different technology in the playback pathway. For example, I recommend an AVR having Class AB type of amplification. This is more musical than Class D. I also include a musical closed box type of subwoofer, actually a RELs subwoofer, because the closed box design is more musical and they can utilize the LFE channel from an AVR and also the Left and Right full range signal so that the bass can include AVR effects and low bass music. I have already explained how Sigma-Delta chips provide accuracy, while R2R transistor technology provides a more natural sound, so I tip in the direction of the Denefrips DACs, at least as an upgrade option. Modern sounding Tube Headphone Amps, Tube Pre-amplifiers and Tube Power Amplifiers also provide a neutral but natural sound, a more holographic lifelike sound, yet rather than having all I suggest a Tube based Headphone Amp and Pre-Amp and a Solid State Power Amp. Therefore, My idea of integration is to utilize these typologies in the playback paths: Sigma-Delta or R2R, Tube technology, and Solid State (Transistor based).

Notes:

(a) I prefer a computer for music streaming, suggesting the best alternative an an iPad, these being the most versatile and straight forward when wanting bit-perfect playback. By avoiding integrated streamer DACs you can choose whichever streaming service you prefer and be able to swap services as you wish. Also, using your computer or iPad makes storage of purchased music relatively simple.

(b) I prefer to use the preamp in the AVR, because is is a simpler integration, compared to an external preamplifier with a 2-way Amp-to-two-speakers switch, though it likely depends on your existing gear as to which approach to take. In the first case, I can add a Power Amp to the main Left and Right output channels of the AVR (as long as I have pre-amp outputs on the AVR). In the case where I have a dedicated preamplifier, I either need another set of main left and right speakers, or a Solid State Power amp into a 2-way Amp-to-two-speakers switch.

(c) I am preferring a RELs subwoofer to prioritize a musical sound (realized through a closed box and Hi Level as well as LFE connections). Another person may prefer a room shaking ported subwoofer design, such as the SVS range.

(d) I am not trying to use AVR Room calibration, preferring to avoid that digital processing step and single listening position limitation. Instead, I rely on Room Acoustic Treatment. This makes for a more analogue sound potential, and shared room listening experience. Room calibration can also be applied through EQ to a separate unit in a stereo system (See the following EQ chapter), but also in that case I similarly want to avoid it.

Music shaped by parallel system to AVR; sharing the main Left and Right Speakers. Here, I repeat the more extended system introduced at the start of this chapter:

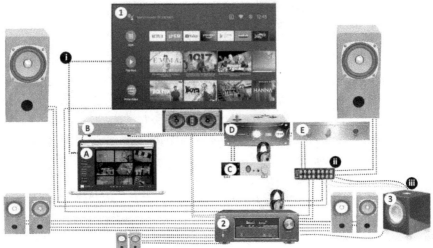

(i), (ii), and (iii) are interconnection points between the Smart TV AVR setup and the Hi-Fi setup. (i) is where the computer uses the Smart TV as an external screen, (ii) is a 2-way-Amp-to-2-speaker switch, so you can switch to use the main Left and Right speakers with both systems. (iii) shows a RELs subwoofer connecting to the LFE on the AVR (for movies played through the AVR), and also using its high level connections to the left and right speakers, at the switch, for both systems.

    1,2, and 3 are the main devices in the Movie/TV/Netflix to AVI Surround Sound system.

(A) to (E) are the Devices in the Hi-Fi System. (A) is a computer used for streaming and other work. Naturally, a router connects to the computer, additionally the computer likely stores some music files on an external drive. The computer on-connects to a DAC. Of course, other devices can be used for streaming, and that might be preferable for you if you do not want to utilize a TV as an external monitor. (B) is a DAC, Sigma-Delta types being the most common, R2R DACs being slightly better. In Streamer DACs there is an (A-B) combination. (C) is a Headphone Amp. (D) is a dedicated Tube Preamplifier to enhance the audio quality to the main Left and Right speakers. It can easily be added between a DAC and Power Amp as an upgrade. You can also connect in other input devices like a CD or Phono player. While Integrated Amps are more common, combining (C-D-E), you get better sound and versatility by separating them out, and, the headphone jack in an Integrated Amp will be inferior. (E) is a Solid State Power Amp; thus, (B-C-D) is imagining Sigma-Delta or R2R, Tube, and Solid State typologies being utilized for their respective benefits.

Using a Tube Power amp to same speakers as an AVR uses.
There are various possible configurations. Here I show a Tube Power Amp attached after a Pre-amp, but the Pre-amp can be switched , in this case between a dedicated Pre-amp and the Pre-amp in the AVR. In this case the Power Amp needs to be a dedicated one (not an integrated one). Some integrated tube amps have a switch for Power Amp only mode (which might be the case here). This option is only possible if you have an AVR that has pre-out connections, usually only available on higher tier AVRs:

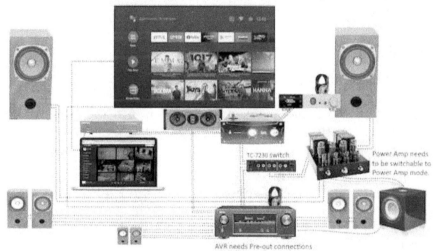

Power Amp needs to be switchable to Power Amp mode.

AVR needs Pre-out connections

I HAVE NOT YET EXPLAINED THE SWITCH IN THE SYSTEM. Normally, it is not considered a good idea to add anything between your Power Amp and speakers, but you can add a quality **2-way amp-to-speaker switch** without a negative sound loss if you pay attention to the connection quality in and out of the switch (see the sections on Screw Down Connectors and Soldering Connections in Chapter 8: Cables and everything related to wire.

The 2-way amp-to-speaker switch I like is the Beresford TC-7210 Selector or TC-7220, see http://www.homehifi.co.uk/S/selector-phono.htm:

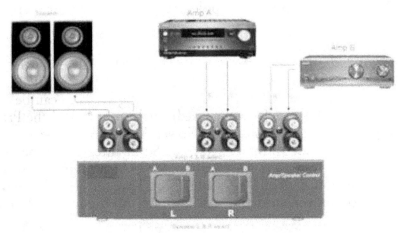

Connection notes:

i) It is best not to try and connect a dedicated Pre-amplifier or Integrated Tube Amp to an AVR, before or after, because of the mixed tones and issue of running a pre-amp into another preamp, thus adding noise, and the inconvenience for TV or movies of having to wait for a Tube Amp to warm up.

ii) The PRE OUT on an AVR, used to on-connect a better quality Power Amp, is a common approach, and I think addign a Tube amp is a good idea. Mind you, the entry to medium level AVRs may not have Pre-outs, and the higher spec AVRs that do, then, also have better amplification.

iii) Another way is to avoid integration. In that case, to run two separate systems: one for stereo hi-fi, the other for TV/AVR, it would require an additional set of Left and Right speakers, which is likely to be over imposing, but a choice you may prefer. In that case, you could add in-wall speakers for the main Left and Right of the AVR instead of an over imposing second set of speakers, remembering the audio for Left and Right Speakers only carry part of the sound of a Surround Sound or Dolby Atmos system, and would be fine for movies and TV. Also, Left and Right in-wall speakers either side of a TV might be easy enough to wire up with the speaker cables only needing to be behind the dry wall down to floor level.

iv) The TV connections works, by an HDMI (arc) connection to my AVR HDMI 2 (arc) input. My TV is a Panasonic Ultra HD TV TH-49GX740Z. From my computer (I use a laptop), because I only have a Display Port, I connect out via its Display Port to an HDMI converter and on to the TV HDMI 1 input. The display settings on the computer are set as extended, scale and layout 100%, Display Resolution: 3840 x 2160, which is the same display Screen Resolution of my 4K TV. For the computer Mouse, from Settings>Mouse>Adjust Mouse and Cursor Size, I increased the size, so when I move it to the extended 4K screen it is not too small. I also increase the cursor speed by about 30%. Since the HDMI from the computer goes into the TV HDMI 1, that is what I select on the TV remote INPUT to view the TV as an extended computer screen. To view the Smart TV, I set the TV Remote input to AV. Since the sound goes out of the HDMI2 (arc) to the AVR, when I want to get into the AVR settings, I select HDMI 2 on the TV remote, then on the AVR remote, I select Setup.

v) The Sound Output on my computer is selected to my my Music Player, which in turn selects my DAC driver. I sometimes play music on my computer either with a Media player called JRiver, and from JRiver I set the Audio output to the DAC driver. JRiver has a member music sharing server via Cloudplay which provides some good quality music to explore. JRiver can also play YouTube and access Amazon Music. I also Use the Amarra Luxe media player for accessing Tidal (though I use the Tidal Account just a little in holiday periods – any music I discover I like I can buy online, or buy some super cheap CDs and rip the music from them, or buy the higher res on-line.

DSD – how it fits in

Many audiophiles believe DSD is the best viscerally analog sounding digital file type. Although a direct comparison between DSD and PCM is not plausible, DSD in some respects is comparable to a 20-bit/96kHz PCM file. So you might say it is hard to tell the difference between high sample rate PCM files, which you get through the top streaming providers and DSD, which you get though SACD and by buying files or albums from online retailers. DSD is better, but you do need good gear and room acoustics to realize it. That gear will be the higher priced headphones, and probably utilizing R2R/ Multibit and/or FPGA typology in your DAC, and/or Tube pre-amplification and/or tube amplification. That quality of gear, edging into Hi-end will also improve the clarity of high sample rate PCM like 192kHz.

Hearing the improvement above 96kHz and DSD24 is also about the quality of the recorded material in the first place. Note here that where quality recordings are available, such as master tape recordings that were used to produce SACD, those particular albums will also probably be available at 192kHz PCM from streaming services like Tidal (Tidal Masters), Qobuz and Apple Music. Most SSACDs that were remastered are of jazz, blues, classical or classic rock/pop titles.

SUPER AUDIO CD

In more modern times, most of what is recorded for SACD or DSD is jazz, blues, classical or solo artists, not classic rock/pop titles.

So then, it is probably not worth pursuing DSD, unless you are also going to chase Hi-end gear, such as available from PS audio, see https://www.psaudio.com/collections/. Or, unless you already have a SACD player and a SACD library that you can also rip to files, and get an even better rendition through an external DAC than the DAC in the SACD player.

But, I am not a high-end audiophile. I don't want, at this point, to move to high-end gear, like the many-thousand-dollar-commitment DirectStream DAC MK2 from PS Audio, see https://www.psaudio.com/products/directstream-dac-mk2. Back down to earth for those of us on a budget, however, is to try out some DSD files by buying a few you really like, while perhaps upgrading your room acoustics and gear. If you like the music available in DSD, sometimes by lesser known artists, but spectacularly recorded, you might catch the bug.

Since you don't stream DSD, the files will be stored locally. The files are too large to stream. A dozen albums in DSD256 will take up about half of a Phone's 256GB storage space. For at home, other methods of storage include expansion drives and NAS file servers. I use a 2 TB Seagate Expansion Portable Hard drive for all my data, which also covers me in case of a computer crash.

You can buy DSD music from these reputable sellers:
Native DSD: www.nativedsd.com/
Channel Classics: www.channelclassics.com/
Blue Cost Records: bluecoastrecords.com/
Forward Studios: www.forwardstudios.it/en-gb/store/Music-c95566325
Pro Studio Masters: www.prostudiomasters.com/ [All types of formats]
Chasing the Dragon: chasingthedragon.co.uk/shop.html

HDtracks: www.hdtracks.com/ [All types of formats]

As mentioned, SACD owners might also rip the SACDs to DSD64 files for longevity and mobile playback, and better sound with a modern DAC/AMP than you get from the DAC inside of a SACD player. Actually, you could get a second hand SACD player and build up your SACD collection, such as purchased from https://store.acousticsounds.com/c/4/SACDs. And, here is how to get into ripping SACD to DSD files: https://www.psaudio.com/blogs/copper/down-the-rabbit-hole-of-sacd-ripping-and-dsd-extraction.

SUMMARY

WE BEGAN by understanding the various kinds of ear buds and headphones, focusing our judgments on the tonal and acoustic benefits of different types, which included aspects such as Bluetooth and Atmos. When considering DAC/AMPs in the playback path, I sorted the DAC ideas by chip and circuitry typologies as one way to appreciate sound signature differences, moving from Sigma-Delta, to FPGA, to R2R/Multi-bit, Solid State, and to Tubes.

The advice I offered showed the advantage of using Apple devices for serving music, but then I suggested using a computer, and I suggested using container media player software like Roon, Audirvana, or Amarra, because they go that extra step in ensuring bit-perfect hi-res playback and sound quality across operating systems, and they access locally stored files, which may include DSD files. Then I proceeded to discuss the various potential gear in a Hi-Fi and TV-AVR integrated system. We finished off the chapter with some discussion on how the DSD format fits into things.

CHAPTER 8
CABLES AND EVERYTHING RELATED TO WIRE

"Racing is a constant search for the weakest link", by Duane Bailey.

IT IS QUITE POSSIBLY your wires and cables, and/or the connections, are the weakest link in your audio chain, so this chapter explores what we need to know, to plug any potential gaps. We research power cords, cabling, and wiring connections. At any point there can be a weak link.

RECORDING: MICROPHONE TO DIGITAL INTERFACE

LIKE SPEAKER WIRES, THE CABLES FROM dynamic and passive ribbon mics pass an analogue signal until digitized in a digital interface. While a signal is analog, and especially if weak as from a ribbon microphone or vinyl record player/phono, the audio quality is more so tonally affected by the cabling, I believe.

For my Ribbon Microphone, I use a VOVOX Sonorus Direct S cable (1meter long), see vovox.com/collections/xlr-kabel/products/sonorus-direct-s-pair. I bought it here, www.thomann.de/gb/vovox_sonorus_direct_s100_xlrxlr.htm. It connects from my microphone to my Grace m101 ribbon preamp. Then to connect my preamp to my digital interface. VOVOX Sonorus Direct S cables are made form high-grade solid core copper with refined crystal structure with natural fibers netting around every single conductor wire. I do not know what AWG it is. They are not shielded, and they have Neutrik XLR connectors with gold plated contacts. I presume the soldering to the connector is top quality. I found this cable sounded quite a bit better than my Mogami Gold Studio. I think the difference shows up more because ribbon microphone have a very weak signal until they are per-amplified up. It is interesting to me that without a shield they sound better, though, because of that I have to make sure the cable is in clean air, not parallel at all to other cables, and well away from power cords.

I also tried the Sommer Epilogue QuadCore (1meter long), see shop.sommercable.com/en/Cable/HiFi-Home/NF-Phonokabel-Stereo-Paar-Epilogue-QuadCore-HighEnd-EPB1.html, again between my Ribbon Microphone and Pre-amplifier. This one was better than Mogami, but not enough to shout about it. The Sommer Epilogue QuadCore uses four core wires with carbon conductor smoothing. The four wires are shielded with copper mesh, and there are gold-plated connectors. I use a these between my ribbon pre-amp and Digital Interface, and I use it for my condenser Shotgun microphone to the Digital Interface:

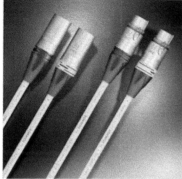

I think you might also go to this sort of level if you are connecting a vinyl/phono player to a Phono pre-amp.

 ## OPTICAL CABLES

SOMETIMES YOU MIGHT BE USING OPTICAL CABLES when connecting up gear. Optical cables will pass up to 5.1 channels of surround sound, but not more modern formats like Dolby TrueHD and Dolby Atmos. For example, I have used an optical cable from my BabyFace Pro digital interface (DI), which I also use as a DAC, to my AVR. In that path, I am only interested in Stereo, so I select "SPDIF" for stereo, then "PCM" as the digital format.

Optical cables can pass up to 24-bit/192kHz if the sender and receiver can do it, see ayrn.io/toslink-optical-connections-support-24-192khz-audio/. I don't think it is scientific, but I did get a glass fiber optic type – even though I think it is overkill (unnecessary). It cannot hurt. Lifatec glass optical cables get good reviews, see http://www.lifatec.com/toslink2.html:

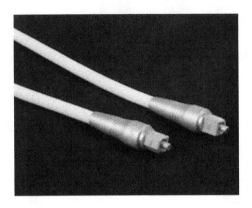

 **INTERCONNECTS**

INTERCONNECT CABLES (sometimes called line level cables) connect gear up. For monitoring when working with audio, between my DAC and Powered Speakers, I use Mogami Gold Studio XLR see, mogamicable.com/category/gold/. In this case, the XLR is a balanced cable. Balanced cables work by sending two signals down the cable and rejecting any difference between the two. For this to work you need XLR connections at each end. In terms of quality, Mogami Gold is a good standard. Here is a video by Kettner Creative discussing it, see https://youtu.be/jWJ10EWQnTI

OFC copper, is an excellent conductor. OFC means Oxygen Free Copper and the manufacturing process, which forms many small crystals. It boasts better sound in Interconnect cables. By conductor, I mean electrons pass through it really easily so as to neutrally transfer the original sound. Sort of. Different purity levels and crystal structure has an effect on sound quality, and how the cable is built in terms of core wire versus wire strands, the shielding, the type of insulation, and shielding, and how good the connectors are and how well they are soldered. Generally, when you pay more, more effort has gone into these aspects.

IN TERMS OF COPPER VERSUS SILVER, PURE SILVER, 4N (99.99%) is a better conductor, 7% better, than pure copper. The 7% means if you have the same purity of silver and copper, and they are the same lengths, and if the silver was about 7% thinner, the exact same electrically wave would pass through it, and so the same frequency curve, but does it sound the same? No. Silver actually sounds a little brighter than copper and so is mostly not prefered by the industry or audiophiles. Bright means it sounds a bit clangy. Fast, yes, but clangy. Let's compare this to copper.

What I have said, so far, is debatable, because the reasonable quality cables generally measure exactly the same in terms of frequency curve as more expensive ones with more build complexity. That is used as an argument that audiophiles are only imagining different sound signatures after paying more for their cables. However, a same frequency curve measured going in one end and out of the other of a cable only shows that the volume for all frequencies are the same, not anything about tone.

Watch this recent video by Jaap Veenstra of Alpha Audio, dated 4 April, 2024, on extensive measurement process of different Interlink cables, see https://youtu.be/8wVnURAckLI. This shows how, yes, there are measurements that account for some differences in cable sound signatures. Spectral decay is how fast or slow the many sounds in music and vocals fade away/decay and in particular we are looking, below, at the harmonics in that spectral decay. Secondly, roll-off, where energy rolls-off after a pulse of sound. In the spectral displays, below, on the left we see a slightly slower roll-off with more visual fizz/harmonic energy, thus a slightly more energetic, airy and fresher sound. Correspondingly, in the Spectral decays on the right we see a bit quicker roll-off, with less visual fizz/harmonic energy, thus a slightly drier/warmer sound. Here is the full article by Jaap Veenstra of Alpha Audio, see https://www.alpha-audio.net/review/interlinks-dont-do-anything-or-do-they-32-rca-cables-analyzed/.

Here is a comparison of four of the cables that were tested:

Van den Hul MC Silver IT MkIII Balanced
Price: $2200, €2000
Tone: Lifelike voices. Accurate. Piano also feels incredibly real.
Spectral decay slightly slower roll-off with more visual fizz/harmonic energy, thus more slightly more energetic, airy and fresher sounding.

Grim SQM:
Price: $845 €780
Tone: Lots of detail audible. Both in instruments and vocals. However, balance between tenor and soprano is not right.
Spectral decay roll-off is a bit quicker, with less visual fizz/harmonic energy, thus slightly drier/warmer sounding.

Material: Matched Crystal (MC) Oxygen Free Copper (OFC), thick silver coated, inside teflon coating, inside proprietary Hullflex EHD outer jacket.

Material: Two twisted OFC copper pairs (presumed), carbon doted sheaths coated, inside a carbon dotted sheath, inside a braided shield.

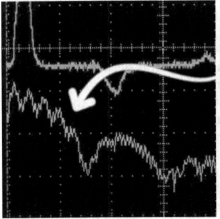 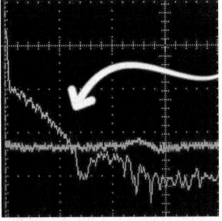

QED 40i:
Price: $105 €99
Tone: Voices sound natural. Harp stands out on this cable. Saxophone is clear (less warmth). Melody is easy to follow.

Mogami 2534:
Price: $37 €35
Tone: Full body in the Saxophone. Much detail audible without being restless.
Spectral decay roll-off is a bit

Spectral decay slightly slower roll-off with more visual fizz/harmonic energy, thus slightly more energetic, airy and fresher sounding.

quicker, with less visual fizz/harmonic energy, thus slightly drier/warmer sounding.

Material: One OFC copper, one silver plated OFC copper solid wires foamed polyethylene dielectric coating, inside braided shield, in PVC outer jacket.

Material: Two twisted OFC copper pairs, PVC coated wires, inside a bare copper shield, inside a polyethylene outer jacket.

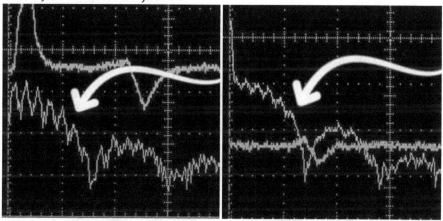

Then looking at the materials, we can see the general differences. I have included, below, those interconnects tested in the Alpha Audio project, that are below $300 for a 2 meter Interconnect, including the notes on vocal and instrument tone, and including the cable build.

Since the testing was done on a reference setup, that means you are hearing the cables differences, not the gear. If your gear is brighter or warmer, you can take that into account, just as you can take into account your own listening preferences. If you are not sure of the difference between warm and bright, listen to this discussion and demonstration of pianos by Family Piano Co, https://youtu.be/KMd7wD_YpR4. For singers think of the warmer tones of Beyoncé, Johnny Cash, Reneé Fleming and Kurt

Moll (Opera), Marin Mazzie and Norm Lewis (Broadway). A warm to neutral tone is more often found in Acoustic, Jazz, Classic Rock, and Classical music. A neutral and detailed tone tends to be found in Pop, Rap, RnB. And, naturally, brighter singers include Miley Cyrus, Justin Bieber Dawn Upshaw and Robert Weede (Opera), Bernadette Peters and Michael Crawford (Broadway). Modern music tends to be mixed on the brighter side to sound for forceful on mobile devices and cheaper (muffled) gear, see this explanation by Steve Guttenberg Audiophiliac, https://youtu.be/bipYWEWNc6w *. That explains why seeking well recorded music is a goal for audiophiles who like to listen to music on better quality systems.

Knowing that, if your gear is on the bright side, such as sigma-delta chips in your DAC and solid state amplification, you probably want warmer interconnects to achieve a neutral tone, or differently according to your music and tone preferences. If your gear is in the warmer side, such as Tube amplification, you may try slightly brighter cables to achieve a neutral tone, then vary the actual tubes according to your music and tone preferences. I hope this table helps as a purchasing reference:

Cable	2m cost	Tone of vocals and instruments	Overall sound	build
Audioquest Red River	$298 €279	fine "weight" throughout.	nice sound. Fine body.	PSC (smooth surface copper) solid core
Chord Comany C-Line	$122 €115	Whiff of warmth (strange, isn't it?). Fine body in vocals.	Easy listening. Calm sound. Overall nothing distracting.	OFC, stranded.
Grimm TPR	$106 €100	Lots of nuance and detail in voice. Lots of body in piano.	Fast cable. Lots of layering in stereo image.	OFC, stranded (twisted pair)

210

Kimber Tonik	$233 €219	Long reverb. Clear imaging. Good focus.	Light and transparent. Good imaging. You are on the edge of your seat.	'ultrapure' OFC
Mogami 2534	$37 €35	Full body in the Saxophone. Much detail audible without being restless.	Fine cable in terms of balance. A bit smaller in terms of image though.	OFC stranded – quad core cable.
QED 40i	$105 €99	Voices sound natural. Harp stands out on this cable. Saxophone is clear (less warmth). Melody is easy to follow.	In terms of timing, a very correct cable. Sound is mid-focused.	OFC, solid core.
Supra DAC	$148 €139	Slightly noisy cymbals. Saxophone a bit thin, but piano is rich in sound.	Careful and cautious sound. Measured and precise. Timing is good.	Stranded, OFC 5N copper
	$298 €279	fine "weight" throughout.	nice sound. Fine body.	PSC (smooth surface copper) solid core

And, there is more testing to come, if you are interested, see this article by Jaap Veenstra of Alpha Audio, which I presume will be updated as the research progresses, see https://www.alpha-audio.net/review/interlinks-dont-do-anything-or-do-they-32-rca-cables-analyzed/.

* Related to the issue of overly bright recordings in modern times, I explained in chapter 6 how I use the VST plugin, Tube-Tech Classic Channel mk II, from Softube,
for playback on my hi-fi system, placing it as a VST plugin in JRiver Media Player's DSP Studio, usually on the preset, 'JC Hi-Fi Vocal Chain'. To my ears the warmth tames down the overly bright emphasis in many modern recording – yet I still have sparkle and liveliness:

ANOTHER WAY TO ADJUST BETWEEN WARM AND BRIGHT TONE IS TO ADD AN EQUALIZER like the analogue domain **Schiit Loki** (inserted after your DAC), see https://www.schiit.com/products/loki-mini-3, turning down the 2K and 8K for warmth,; turning them up for a brighter punchy experience. See this review by Steve Guttenberg Audiophiliac, https://youtu.be/Hj9lR2NiaPg.

Related to tube amplification, there is **a classical Tube sound which is warmer, while the modern Tube sound is more neutral and transparent** due to improved circuit design and higher quality components (thus benefits less from silver). So, you might use copper interconnects into a Solid State Amplifier, and use only a degree of silver or copper (where the cable typology anyway is a little brighter) for modern tube amplification, and more likely some silver in interconnects for older tube gear.

Your Tube choices takes it from there, according to your speaker type and preference for the aspects of resolution, dynamics, sound quality, sound stage, and character. This tube choice experimentation is nicely illustrated by I Am Mad's Tube rolling experiments with the Pacific Paradise Pre-Amp and full range and 2 or 3-way speakers, see https://youtu.be/XxqtiDKJA0E.

I USE **SILVER COATED COPPER** FOR MY INTERCONNECT FROM MY AVR TO **SUBWOOFER,** being that there are no high frequencies involved, the benefit is on improved speed / tighter bass. My subwoofer has, like most subwoofers, an RCA/Coaxial cable connection for LFE (Low Frequency Effects), sometimes called 'Sub out', whereby the subwoofer accepts the LFE signal from an

Integrated Amp or AVR. This RCA cable needs to be double-shielded. I bought a YYAudio subwoofer interconnect, listed as RCA Cable RCA to RCA Cable Coaxial Audio Cable Subwoofer Cable SPDIF Male Stereo Connector for TV Amplifier Hifi Subwoofer on AliExpress, see https://www.aliexpress.com/item/1005004654922018.html?spm=a2g0o.order_list.order_list_main.15.593818026j03uF:

The copper is 7N OCC, which is high level purity. I went with Silver Coated Copper for the increased speed, in addition to the double shielded design to fend off external electrical interferences.

The same **silver coated copper** approach is taken with a somewhat similarly reputable subwoofer cable, the Cable Co, see https://www.thecableco.com/cables/subwoofer-cables/the-sub-hybrid-subwoofer-cable.html. In that case they use silver coated stranded conductors. Not all though. Here without any silver, a well regarded copper subwoofer interconnect is BJC LC-1 Subwoofer Cable from Blue Jeans Cable, the BJC LC-1 Subwoofer Cable, with double braided shield, see https://www.bluejeanscable.com/store/subwoofer/index.htm. If you read their data on design of the cable, it is all about shielding and keeping the capacitance low for excellent dynamics and timing.

I use **a pure copper Interconnect** for my REL's **subwoofer High level connection** from the main Left and Right terminals of my AVR. The High Level interconnect accepts the full signal (not just LFE channel signal) as a supplement for stereo hi-fi playback, which it controls via crossover and volume settings on the subwoofer. The high level connection and the RCA LFE connection work at the same time, so you get low end stereo supplement and LFE effects. That is how RELs do it. If you buy a RELs subwoofer, it comes with its special high level connector, and instructions for connecting up depending on the amplifier class, AB or D class. The black wire acts as a grounding reference. I bought this REL 3 Wire Sub Speaker Cable upgrade with a Speakon connector from Design a Cable, see https://www.designacable.com/rel-mj-acoustics-3-wire-sub-speaker-cable-neutrik-male-xlr-to-bare-end-subwoofer.html

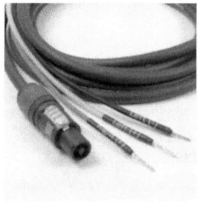

SPEAKER WIRE

WIRE QUALITY PARTICULARLY MATTERS particularly when the signal is analogue. The first thing to understand is that copper, then pure OFC copper, is the most popular choice for speaker wire, and for good reason - it is the same material used in the production process and it is the best conductor, bar silver (which is very expensive, and mostly not used in the production process). Nevertheless, some silver options are discussed below.

Then there are other considerations, such as braided wire, more so for speaker cables, which helps prevent external noise interference from entering the wire. Here, various unbraided speaker wires are tested as antennas, showing how wire can easily pick up external electrical interference and thus add noise to whatever electrical current is passing through it. Braiding overcomes this, see this test by GR-Research, youtu.be/DC0s6KqQz3g.

Here is a link on how to connect this sort of wire, see E-Z'S https://ezeescrossovers.com/braided-speaker-cable-assembly.

Less bulky, less costly, and already connected, Steve Huff, a well known reviewer, recommends the 14 AWG Micca speaker cables, see https://youtu.be/44n1AEQO6nY?t=315. On eBay, search under, "Micca 14 Gauge Pure Copper Speaker Wire". You can get them in 6 or 12 foot pairs:

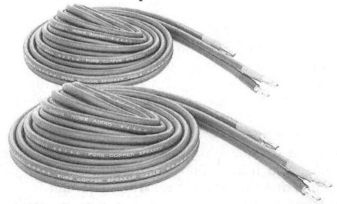

Yes, there is a potential issue of dual parallel wires of acting like aerials, governing cheaper constructions like these rely on not the best PVC coating and no shielding, so using the 6 foot rather than 12 foot lengths might be the way to go with this speaker wire, and/or keep each strand of wire apart by a centimeter (perhaps with the help of some Velcro.

WITHOUT GOING INTO TECHNICAL DETAIL, here are key aspects of speaker cable design to look out for:
- pure copper, oxygen free (OFC). OFC limits copy oxide building up between copper crystals in the cable.
- multi-strand.
- braiding.
- Teflon, PP (polypropylene) or PE (Polyethylene) over the wire rather than PVC, (Though PVC jacket as a final layer is good for strengthening.
- cotton core or as part of the internal insulation.

FOR SPEAKER WIRE THICKNESS, thick wire (12 or 14 gauge AWG) is recommended for long runs, high power applications, and low-impedance speakers (4 or 6 ohms). For relatively short runs (less than 50 feet) to 8 ohm speakers, 16 AWG wire is a good match. When AWG numbers go up, it means the wire is thinner. The shorter the distance the thinner you want; the longer the distance the thicker you want.

If you are wiring up inside a speaker between the crossover to speaker drivers and speaker wires to binding posts, use thinner quality wire, 14 or 16 AWG since it is for shorter distances, see this discussion by Paul McGowan of PS Audio, youtu.be/KYLDGe9K9OM.

 **SCREW-DOWN CONNECTORS**

SOME SPEAKER BINDING POSTS have **spring clips**, which clamp down on the wire ends. If you have spring clips, watch out for whiskers of wire – you do not want them touching the wire of another terminal or you will get a short. Spring clips do not offer a really solid connection, and may mean the connection is the weakest link in your audio pathway. They work, but you can do better.

Generally, welded connectors or ones that are clamped down or plugged in are considered a more solid connection. I recommend screw-down spades or banana plugs, and to make a good mechanical cold weld. To achieve a cold weld, where you simply crimp or screw down the wire, the pressure is forcing out surface roughness, so that the metal atoms can jump from one piece to another. Cold welding requires cleaning the metal surfaces to 'perfection'. The oxide layers need to be removed, as does any surface grease and dust (including finger grease).

To do this, I suggest:

(a) If there is wiggle room in the receptor hole, double over the wire, so when pushed in it is a snug fit. In that case you can add extra layers of heat shrink, to get a firm fit in the grommet hole*. We want a non-wiggle wire coming from the connector end.

(b) Spread all the wire strands, and use a little P1200 wet and dry sandpaper, held between your fingers. Wet the sandpaper with 99% Isopropyl Alcohol, which you can get from any electrical store or perhaps home depot, and without touching the wire strands with your fingers, run the sandpaper over the strands, until they take on a shiny appearance. Then, using the sandpaper to grip, twist the wire into a light loom, as is commonly seen, (we usually see people do it with their oily fingers directly).

(c) You can also roll a little sandpaper up, sprayed with 99% Isopropyl Alcohol, and smooth a little the inside of the connector and the tip of the screw, then use a cotton buds, again sprayed with 99% Isopropyl Alcohol, to wipe-clean inside it.

(d) Then, screw down the screws on the wire, tight, using a little body weight **, checking it is solidly in place with a reasonable push and pull test.

* As per NASA policy 13.4.5 "When an insulated wire diameter is smaller than the grommet hole, the wire insulation diameter shall be increased by using heat shrinkable sleeving," see https://nepp.nasa.gov/files/27631/nstd87394a.pdf.

** screw tightness force should be about 60% of the tensile strength of the wire being screwed down, which is a fair amount for speaker wire - put some light body weight into it when screwing it down, but not so much as to ruin the screw seating. See note under Table 13-1 p57 of the NASA policy, in the above linked document. Then a pull test should be no more than 80% of the force used to screw the screw down, see note under Table 13-2. P58 of the same document.

BINDING POSTS: Again, you don't want your speaker binding posts to be the weakest links. I would suggest looking for pure OFC copper, or mixed metal, copper or gold plated, see HiFi collective, www.hificollective.co.uk/components/binding_posts.html.hten it down hard, even with a little twist with some pliers.

SOLDERING CONNECTIONS

SOME CONNECTORS ARE DESIGNED TO BE SOLDERED, more so soldering goes on inside gear and inside-speaker-cabinet connections. Certainly, you can save money by making your own interconnects and speaker cable terminations. For the soldering, the steps in this video, youtu.be/8V-m66BKd9U are:
(a) Slip on some heat shrink. Add more heat shrinking if you need it to stabilize the grommet connection more.
(b) Clean wire and connectors, using the system above (sandpaper and Isopropyl Alcohol).
(c) Tin the wire using Rosin core Tin-led solder (63/37 Pb/Sn), and better still add Mundorf Silver Gold Solder Supreme, see

www.hificollective.co.uk/catalog/mundorf-silver-solder-supreme-100g-reel-p-1488.html. Have a fan pointing to blow any fumes away. Here is a video where GR-Research is using this technique, soldering internal speaker drivers, youtu.be/mVcOWx7hQiQ, and youtu.be/J_E4_CyHA5A. When he does the tinning, notice how he heats one side of the wire with a droplet of solder on the soldering tip, while also feeding the solder onto the other side. What you are going to do is have the silver-gold solder flow into the tin-led solder as a secondary process. The melting point of Mundorf Silver Gold Solder Supreme, is 290°C / 554°F, so set the soldering iron higher, 310-320°C / 622-640°F). Alternatively, you will not go wrong with any of the solder available at hificollective, see https://www.hificollective.co.uk/catalog/-c-61_84_196.html. Cardas Quad Eutectic solder is recommended as being very easy to work with, see https://www.head-fi.org/threads/best-type-of-solder-for-audio-quality.533536/.

A soldering station is a good idea, for setting of the temperature, but not necessary – I have a Hako FX-888D. Place the terminator in a third hand soldering tool. If you use a vice, hold the wire away from the tip so the vice does not disperse the heat of the soldering iron.

(d) clean any flux or splattered plastic from the surface.

(e) heat the heat shrink over the weld.

FOR SOLDERING TERMINALS, you might need to check videos of how to terminate certain wires, for example, for Neotech NES-3005 wire termination with Neotech OFC SK8-B spade connection, by HiFi Collective, youtu.be/zP5HE2WGCcw. Another for Canare 4S11, by Tharbamar (youtu.be/snLExbyIJto).

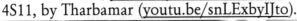
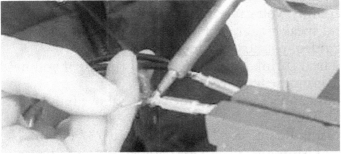

Another, showing four stranded braid with Harmonis Litz wire and Audio Note banana plugs, by HiFi Collective, youtu.be/8V-m66BKd9U). He uses a solder bath, which seems a very good method for pre-tinning the wire, which has melted solder pellets (not silver content) in it. After tinning the wire, he also uses Mundorf Silver Gold Solder, www.hificollective.co.uk/catalog/mundorf-silver-solder-supreme-100g-reel-p-1488.html, which is bests for audio connections, with Technoflux Flux Soldering Liquid, or you could use Aquiflux.

BRAIDING was included in that previous video. Here is another showing more clearly the method, thanks to knot queen, see https://youtu.be/G-7f9F3MF7Q. It's one far-side, under two and back over one; the other far-side, under two and back over one, repeated.

If you are braiding wire, look for an arrow, so the strands face the same arrow direction. It does not matter yet, but once you start using the wire the electrodes align in the direction of travel. If all the arrows on the wire are in the same direction iof flow, it will be easy to check the direction if you swap it over to another setup.

INSTEAD OF BRAIDING, you can separate them with masking tape, as explained by Paul McGowan, PS Audio, youtu.be/Y_jkLCyJFPo. It does not look as good, but it is effective.

ETHERNET CABLES are signal wires, so, depending on how busy your nest of wires is behind your gear, it most likely will pay to have shielded ethernet cables> Supra ethernet cables are often recommended, I bought these ATAudio and Zoerax ethernet cables through AliExpress:

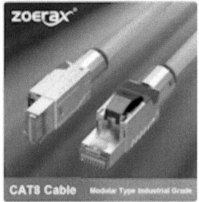

Subwoofer interconnects are also signal cables, and so it is best if they are shielded. My choice was explained above in the section discussing Silver.

HISS AND HUM FROM YOUR POWER SUPPLY

ALL YOUR RECORDING GEAR SHOULD BE PLUGGED INTO ONE WALL SOCKET or CIRCUIT, thus avoiding AC ground loop. I use a Furman conditioner for my recording setup, www.furmanpower.com/power-sequencing/. For your hi-fi gear, just use a good power board into one socket, explained more below.

I have the Furman PL-8C E, which I use for my recording gear, see https://www.furmanpower.com/product/10a-power-conditioner-w-lights-230v/.

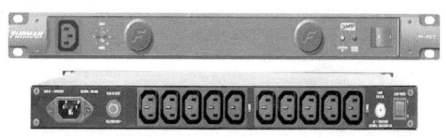

Or, more upmarket, I would recommend anything from Furutech, check out https://www.furutech.com/products/power-distributors-filters/

I decided against a power conditioner for Hi-Fi playback, because although in trying it, it sounds cleaner, it did reduces top end overtones, such as when a guitar is plucked (because of added impedance). See this discussion by Paul McGowan of PS Audio, https://youtu.be/RYXlWp3p9qA. Instead, for my Hi-Fi setup, I use a surge protected powerboard. I presently have a 12 outlet Jackson RAC 1200, see https://jacksonpower.com.au/products/rack-mounted-surge-powerboard-12-outlet:

If you have the budget, you cannot go wrong with Furutech power distributors, for example the e-TP86(G), see https://www.furutech.com/2017/05/27/19764/:

Keeping an eye out for second hand Furutech power distributors on eBay might be well rewarded, remembering you need the right power rating for your country.

FOR A SOLID CONNECTION TO YOUR WALLS POWER SUPPLY, I suggest upgrading the wall socket / receptacle that you plug your into your surge protected powerboard or conditioner into. Or, replace it, if it is a bit old. There are plenty of YouTube videos showing how to do it. The same goes for any other power socket you are using for your audio gear – trusting that it is on the same circuit. If you can, get a hospital grade or commercial spec grade power socket or one from Furutech or elsewhere, for the benefit of the better metals and grip connection. Or, simply replace the one you have if it is a bit old (minimizing oxidization issues), and while reconnecting, clean up the wire end – just snip off the old bare wire and strip back a new section. Sand the exposed wire lightly with some very light sandpaper, such as P1200 wet and dry sandpaper, wetting the sandpaper with 99% Isopropyl Alcohol.

You might come across advice to connect power amps directly to its own power socket. I think this is a hi-end idea where you get an electrician to run a dedicated line from your circuit breaker box.

THE REASON IT IS BETTER to plug all your audio equipment into one wall power socket, or two wall sockets as long as they are on the same circuit. This is to avoid the risk of something called AC ground loop. This is when two pieces of audio equipment are connected together, then each is plugged into different circuits, with each circuit providing slight differences resistances in their ground wire, thus causing small voltage differences in your wires (via the circular ground connection). You'll get a 50 Hz or 60 Hz hum, depending on what country you are in.

UPGRADING TO SHIELDED POWER CABLES, is a good anti-noise anti-static insurance. Sometimes you hear the argument against upgrading power cables as being of no affect since you already have great lengths of pretty poor standard power wire (in terms of audio) running through your walls, so what difference could adding a meter or so of real quality wire at the end of it make? For just a few devices, sure, but that is overlooking a number of things. One issues is noise cross contamination from power cords to audio signal wires. When multiple power wires run together behind your stack of audio gear, they can easily run alongside speaker wires and audio signal wires. Analog signal wire can pick up hum and/or electromagnetic interference (EMI)) from power cords, if not crossed at right angles, Therefore, just as electricians do not run power wire together with signal wires behind the walls, we should keep power cords separate from signal wires. That can be difficult as your wiring starts to look like a birds nest behind your gear. Having shielded power cables solves the proximity problem, so even if they rub against signal wires it should not cause issues. Secondly, your connections are often the weakest link because of oxidization / resistance build up, so when you upgrade power cords, you will be also upgrading the plugs which should have better metals and perhaps be gold plated. Here is my nest of cords and wires, behind my audio gear:

I have a shielded power cables from my wall socket to the power board, and from the power board to my AVR, to my subwoofer, and even to my printer (because my printer power cord runs close to my speaker cables). I added a few spare bits of acoustic material to ensure the speaker wires are not resting on my powerboard, and a piece at the top to separate the AVR power plug area from the optical and HDMI cables.

Here is what shielded power cord looks like:

Here, A British Audiophile touches on this noise issue between power and signal wire, explaining what a shielded power cable is, see youtu.be/Dtb88_hbCFQ?t=824, and here he shows how you can construct them, see https://youtu.be/gfVYfC6tmcc. The link he provides for DIY Main power lead sets is https://www.mcru.co.uk/product-category/diy-componants/diy-mains-power-lead-sets/.

FOR COMPUTER AND TV CABLES that have 320-C7 2 Pin figure-8 connectors, there is no ground wire in them, so be tidy and physically separate them from analog signal wires.

YOU MIGHT COME ACROSS ADVICE TO CONNECT POWER AMPS DIRECTLY TO ITS OWN POWER SOCKET. I think this is a hi-end idea where you get an electrician to run a dedicated line from your circuit breaker box.

YOU CAN OF COURSE BUY PRE-WIRED SHIELDED POWER CABLES, if you are happy to pay more, or save by sourcing from China through eBay or AliExpress. Either way, to keep things tidy, use cables no longer than you need them.

As mentioned, THE POWER SOCKET OR CIRCUIT YOU ARE PLUGGED INTO should not have fridges and air conditioners on the same circuit. It is usually the case that you can to use two power sockets in the same room because that will be on the same circuit. If you look at the power board for your house or apartment, you will see a number of power fuses (which are for separate circuits).

Probably there is one circuit for your stove, another for your kitchen, and another for your other rooms. You can test it by pulling, or switching off each fuse at your fuse box, one at a time, and seeing what power sockets become dead. If, for example, you have to have an air conditioner on the same circuit, that will probably add hum to your audio gear. In that case, is there another room without an air conditioner or whiteware appliance that you can run a main lead through the wall to? If all circuits have issues, the best solution is to get an electrician to add a new dedicated power point/power circuit for your audio. Otherwise, you will need to use some kind of power conditioner or power regenerator.

HERE IS A PROCEDURE TO TIDY UP YOUR WIRING AND CABLES:
1. Upgrade power cables to shielded cables of the right length, including ethernet cables. Use balanced XLR interconnect cables where possible. Upgrading to braided speaker wire helps.
2. Feed power cords, from a Hi-Fi console area, down to a lower place where you have your surge protected power board, or conditioner. The cable from the power board or conditioner then runs along the floor, or close to it, to one power socket, or more on the same circuit.
3. For any power cords that are not shielded, such as from your TV or computer, plug them into the left and right sides of your power board, or conditioner, thus they are easily to keep away from other cords and analog signal wires and cables.
4. For speaker wires, keep them separate and above from power cords.
5. If you have to cross analog wires and power cables, cross them at right angles.
6. Pay particular attention to keeping low signal analog wires separate, such as for microphones and phono/vinyl record players.
7. Secure cables if needed, using hook and loop Velcro tape/cable ties, adhesive cable clips, and/or packing material.

A SECOND LEVEL OF THIS GROUND LOOP PROBLEM IS DC GROUND LOOPS, THUS NOISE, BETWEEN GEAR. This problem commonly adds hiss (regardless of whether you are connected to one power socket or circuit, or not). Unplug different devices to see if you can find

the location of the noise. I like this video, which covers the range of solutions to ground loop hum issues, by FireWalk, see youtu.be/einxGsiuwso. It covers, the additional ideas to:
- keep signal cables and power cables away from each other (or at least crossed over at 90 degree angles).
- plug everything into one outlet, or circuit (covered above).
- use balanced audio cables, where possible (such XLR Mogami gold studio).
- if there are any light dimmers in your room, try turning them right down and up, to see if it is a source of hiss.
- if you can hear a hum, with your ear to the chassis of an amplifier or conditioner, and it is at a distracting level, try plugging in an ifi DC blocker+, see ifi-audio.com/products/dc-blocker/. DC is what causes the toroid transformers to hum, and these transformers can be in amplifiers and power conditioners.
- if you hear a hum out of your speakers try a GND-Defender plugged into your amp, see ifi-audio.com/products/gnd-defender.
- Use a computer with a SSD. Most are these days.
- If you have speaker hiss that increases when you turn up your amp with no music coming though, the hiss is from your amp. Perhaps it needs warming up or running in if it's new, otherwise upgrade it if it annoys you.
- If there is wiring, hiss, hum, popping, crackling from your speakers it is probably your wire and/or connections. Clean or upgrade connections and/or wire.

BECAUSE COMPUTERS ARE SO ELECTRICALLY NOISY, you might be tempted to consider adding a USB isolator between the computer and USB DAC. Modern quality DACs already filter the USB port inside the DAC and handle jitter stabilization. Audio Science Review examines the real data, see https://www.audiosciencereview.com/forum/index.php?threads/review-and-measurements-of-schiit-wyrd-usb-filter.5717/ and concludes there is no benefit. Equally for other products see this review from Audio Science https://youtu.be/RulAcLrnPkA Basically get a good DAC, such as is recommended in this book and you won't need a USB Isolator.

SUMMARY

WE BEGAN by claiming the advantage of upgrading interconnect and microphone cables, and a number of options, including discussion of copper and silver. We also considered what makes quality speaker wire, not just the wire itself, but the coating and braiding aspects. Not to forget connector types, and then there is soldering - the finer details of it, important for those who want to DIY or upgrade the internal wiring of speaker cabinets. We considered noise; how balanced cables play a role in reducing it, how electrical noise needs to be minimized, through various measures: power conditioners, power sockets and cords, and by problem shooting electrical hiss and hum. After weakest links have been removed you will be able to reach the potential of the rest of your gear.

CHAPTER 9
CLEAN AUDIO

"You are an analog girl, living in a digital world", by Neil Gaiman.

I WILL LAY OUT TWO LISTS IN THIS CHAPTER. One to summarize what is needed to produce quality audio, the other to summarize what is needed to listen to quality audio. Looming large is the distinction between analog and digital, as I provide a history of analog formats on the one hand being overtaken by the digital age, while on the other hand analog maintains its pride of place in hybrid technologies.

Audio starts as analog when it is recorded, until it is saved in a digital format. Then it is turned back into analog format for amplification and playback. What we want to do is maintain the initial analog purity through any digital processing, or as true to it as possible. In the past analog purity was preserved through saving to tape, whereby the record head converted the electrical signal into a magnetic field which created a pattern of magnetization in the magnetic particles of the tape. When played back from that format, even when copied from a master tape, even when transferred to vinyl, many listeners feel the result is more natural, smoother, more coherent, and more musical than digital. The same argument arises in film, where some famous film producers, such as Steven Spielberg, still today, rely on Super 8 cameras, which utilize newer electronics processing technology while retaining analog film stock.

You can either remain a hard core analog audiophile, sticking with vinyl and/or tape as mediums, or like most of us, because of the versatility of digital audio in delivery and playback options, work out how to get the audio quality you want. Getting that quality requires understanding: what quality recording is, the digital file formats audio can be saved in, ensuring your streaming and gear does not down-sample what you buy, and knowing your digital to analog conversion and playback technologies are scaled to task. I call that clean audio.

RECORDING QUALITY AUDIO

HERE, I WILL LIST, on the recording and post recording side, aspects for achieving quality audio:

1. An acoustically treated room makes the world of difference in getting a quality recording (and for listening) because it tames room modes, limits flat wall echo and helps to balance natural reverberation and direct sound.

2. Quality microphones, chosen for purpose, are a huge factor in recording quality audio. Analog technology, like a ribbon or tube microphones, or VST simulation effects applied to the more common condenser microphones adds to quality.

3. The cables and connectors are important, least they become the weakest link, so these should be up to scratch.

4. A quality Digital Interface (DI) with its Analog to Digital Converter (ADC) and its Digital to Analog Converter (DAC) is very important, primarily because of the need to do a clean job of converting analog to digital, and then digital to analog.

5. A DI should be set to record at 24-bit (or better, 32-bit floating), with a 48 kHz sampling rate (or higher*), volume level set to record at an average -20 to -23 dB, peaking in the orange -3 dB to -6 dB, no clipping, so no high peaks going over -1 dB.

24-bit (or better 32-bit floating) is the traditionally preferred recording sweet spot, because the increased dynamic range makes it relatively easy to set an approximately correct recording volume, while avoiding clipping, and keeping the noise floor below the audible range. In this explanation dynamic range means the extra space below and above the recorded sound (Just to confuse things, for playback, dynamic range is often meant as the distance between the lowest and highest sound).

* A higher sampling rate, like 88 kHz, 96 kHz (Hi-Res), even 192 kHz (Hi-Res) is better because of the way ADCs convert analog to digital. Higher sampling rates do not capture sound in an intrinsically better way – even though it is natural to think so – the same audio is recorded in all cases! As the legendary mastering engineer Bob Katz explains, "the benefit is in the way currently designed digital to audio converters (DACs) work".

Another reason to use a higher sampling rate is that the market is increasingly wanting to listen to Hi-Res audio, so that Hi-Res music vendors may even require 96kHz in the future.

Technically, recording above 96 kHz really is not necessary. I suggest making sure your gear and computer can handle 96 kHz, with the number of tracks and effects you want to work with, without latency or slowing down issues. It would be even better (but not necessary) if you can work in 192kHz. If you have the file space, then operate at 192kHz if you can (as a future proof insurance).

For the purposes of recording acoustic or classical material, where you are not working with lots of tracks and your computer is high spec, you may be interested in offering to market a format called DSD (aka SACD format), more so of interest to the audiophile community. The most common approach being to record at 352.8 kHz PCM (aka DXD), for the editing, mixing and mastering, then converting to DSD256 (see Merging+Anubis Premium digital interface in the chapter on 'Microphones, Digital Interfaces, and Plugins'). For this, you would be well served with the Merging's Pyramix Pro version DAW, which can process the DXD and DSD. Pyramix Pro has core plugins, which compliment the Merging+Anubis Premium built in EQ and compression effects in the Merging+Anubis Premium interface. And, with Pyramix Pro you can add any of the compatible third party VS3 plugins, see www.merging.com/products/pyramix/vs3-vst-plugins. One extra cost would be an equivalent to Izotope RX, because that presently won't support a 352.8 kHz sampling rate. You would be needing the CEDAR Audio Retouch 8 for Pyramix 64-Bit Spectral Editor Plug-In, see www.bhphotovideo.com/c/product/1751216-REG/cedar_audio_cp_retouch_retouch_7_unwanted_noise.html. While CEDAR products are expensive, you are getting value for money, since CEDAR has for some time been the industry leader when it comes to noise reduction.

You can also record in native DSD using the Merging+Anubis Premium interface. In that case you cannot edit anything; you have to record perfectly in one take. Just, record enough takes until you get one right from start to finish - no post recording editing is possible in native DSD, but it is the best final result. Just make sure you have applied the other principles covered in this book - acoustic treatment, quality microphone, quality pre-amplification if needed for a ribbon mic, quality cables, and that other cables and wires represent no weakest link. Also in the case of pure DSD, you may want to look at a hardware tube based channel strip.

6. In every case, apart from native DSD, after recording, you should clean up recorded noise floor and sound intrusions, like rustling clothing, with the quazzi-industry standard Izotope RX, see www.izotope.com/en/products/rx.html. An explanation and images demonstrates this, below.

7. During the mixing and mastering stage, use VST plugins

judiciously. There are a few techniques to do this, to minimize plugin created artifacts. One technique is to apply two passes of the same VST plugin, keeping each pass with very slight settings, like 1 dB or 2 dB. Another technique is to carefully monitor and adjust the output volume of each plugin (aka Gain Staging) so the volume level is the same going out of the applied effect as went in. Noise artifacts are inevitably added when applying multiple VST effects, and so, at least for your main effects, using quality 'clean' filters is important, such as those from Fabfilter, see www.fabfilter.com/products, which also have **oversampling settings** to limit adding their own artifacts.

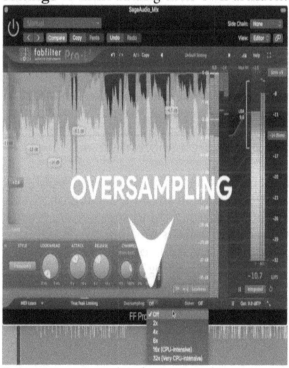

Ref: *www.sageaudio.com/blog/audio-plugins/10-things-you-need-to-know-about-the-fabfilter-pro-l2.php*

Oversampling is effectively doing the same thing as when higher sampling rates are processed. By oversampling as the effect is applied, as little noise as possible is added, or rather removed as the digital processing is underway.

NOISE (ATMOSPHERIC VERSUS DIGITAL ARTIFACTS)

SINCE WE HAVE BEEN DISCUSSING THE NEED TO REDUCE NOISE, we should take a moment to consider what good and bad noise is. This relates to analog versus digital considerations. If you totally remove the noise in an audio track it will sound unnatural - dead. As such something of an emotional connection is lost. The atmosphere in a recording space is part of the sound, and contributing to that atmosphere is any physical aspect in the recording and playback chain. By this I include tube or ribbons in those types of microphones, magnets, and electrical signal too, as it flows through wire and circuitry. As the circuitry lowers in quality, greater amounts of digital noise can be introduced, and in the digitizing process and in the conversion from a digital format back to analog for playback. Digital, scratchy, aliasing noise can be introduced, as it can be during pre-amplification and amplification.

I compare analog reality to walking, where the sound and visceral feeling of feet on the ground adds to the feeling of balance, but not squeaky shoes - that is off-putting.

As we proceed you will see an emphasis on limiting digital artifacts, and removing as much as possible of what inevitably slips into audio, while preserving the analog components of sound – all to maintain an emotional connection to the audio.

HIGH-RES AUDIO

THE ACTUAL AUDIO INFORMATION BEYOND 24-bit/41.1kHz does not increase with higher sampling rates, it is just the free space that increases, but, that free space is utilized by modern DAC processing chips and circuitry, making it possible to have cleaner audio. In the CD rate 16-bit/44.1 kHz we get all we need, but, 24-bit/48 kHz being a bit more dynamic auditory range (free space) than that, provides an opportunity for a DAC to cleanly convert.

HERE IS THE IDEA OF THE HI-RES SWEET SPOT at the merging of the three audio quality parameters: Bit Depth, Sampling Rate, and Bit Rate:

Audio Quality - Parameters

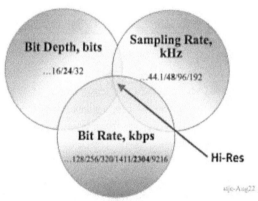

Image by Steve C at https://medium.com/illumination/what-is-hi-res-audio-72a52f19a810

Bit Depth; bits (*the audio resolution detail/ information for each sample/ snapshot*)
Sampling rate; kHz (*how many/ the speed of samples/ snapshots that are taken per second, when digitizing*)
Bit Rate; kbps (*how fast the audio files are able to transfer*)

IF WE HAVE A BIT-DEPTH LOWER THAN 16-BIT we start to hear scratchy noise come through. 16-bit is the point where there is no longer scratchy noise, and so for listening purposes 16-bit provides as much dynamic range as we need – in theory.

The human ear has a dynamic auditory range of up to 120 dB, ranging from a falling autumn leaf up to as close as you can stand to an open air loud speaker:

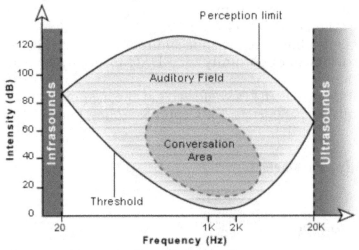

Ref: www.cochlea.org/en/hear/human-auditory-range

CD 16-BIT WITH ITS 96 DB DYNAMIC RANGE, falls a bit short of this, but that will in most cases be a limitation you won't hear, though you probably will be able to hear the different between 16-bit and 24-bit on headphones, more so with classical music that is more inclined to include some super soft sounds. 24-bit audio has a range of 144 dB which definitely covers, with some to spare, the human dynamic auditory range of 120 dB. But, beyond listening, for recording, if we want to apply effects to audio in a DAW, we then definitely need the extra depth of 24-bit as a buffer, so, in the final product, noise is still not evident, and it is even more assured if we use a 32-bit floating bit rate.

Here are three common saved file format Hi-Res settings:
- *Hi-res 24-bit/ **48 kHz** at 2304 kbps, which is recording audio signals accurately up to 22.05 kHz (half the sample rate*, being 2.05 kHz over*

maximum human hearing ability, which is 20 kHz.
*- Hi-res 24-bit/ **96 kHz** at 4608 kbps, which is recording audio signals accurate up to 48 kHz*,*
48 kHz over maximum human hearing/ into ultrasonic - better extra space for the DAC to work with.
*- Hi-res 24-bit/ **192 kHz** at 9216 kbps, which is recording audio signals accurate up to 96 kHz,*
96 kHz over maximum human hearing/ into ultrasonic – even more extra space for the DAC to work with (When listening, you start to need hi-end gear to hear the improvement.

* Let us consider the sampling rate. It might seem weird that we only get half of the sample rate as accurately represented, a fact establish by mathematical analysis and experimentation, to arrive at the Nyquist-Shanan sampling theorem. Above the half way point, the audio is deformed. What we want to accurately represent is the audible range from 20 Hz to 20,000 Hz (20kHz). If we get a bit of clean audio space above that, before entering deformation territory, which in that case starts at 22.05 kHz, then between 20kHz and 22.05kHZ we can apply a roll off filter, to filter out high frequency noise.

 ## FILTERS USED IN THE ADC PROCESS, AND CHIPS

FILTERS PLAY A BIG PART IN PREVENTING NOISE from getting into audio, and de-noising it after the fact. Firstly, when a DAC receives an **Analog Input Signal,** it runs a front end **Analog Low Pass filter,** just before the **Analog to Digital Conversion** takes place:

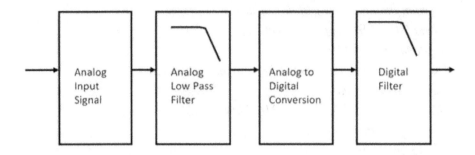

Secondly, since the process of Analog to Digital Conversion introduces some quantization noise (digitizing artifact errors), a **Digital Filter (or filters)** is/are applied to reduce that.

Thirdly, in addition to the filters, additional algorithms do any number of noise shaping tricks. To get some idea of the noise shaping tricks we can look at the Izotope RX Spectral De-noise module>Advanced settings, the overall effect being to mask and shift quantization and/or aliased noise into the ultrasonic frequencies; above human hearing range. Higher sampling rates provide more room for these techniques to be applied accurately. Here are a few noise shaping tricks:

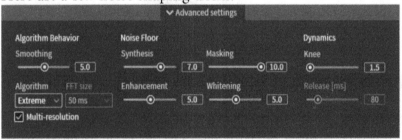

This link explains these and other noise reduction and noise shaping filter processes: downloads.izotope.com/docs/rx6/34-spectral-de-noise/index.html. The actual Izotope RX Spectral De-noise module, allows an audio engineer to apply additional noise reduction after the initial digitizing process. For the audio listener, that same process is underway inside our DACs as they turn digital content into an analog signal.

SINCE WE ARE CONSIDERING NOISE REDUCTION, we should pay attention to the actual chips and technology used in the processing. Delta-sigma chips are the most common and clinically accurate, R2R being a bit more expensive and truer to analog, multi-bit a digital version of R2R, somewhat of a hybrid between Delta-sigma and R2R, and FPGA-based processing, which is a bit more expensive again, and aims to be process audio more efficiently with less noise; truer to analog. These technologies were discussed along with examples of DACs in the chapter on 'Headphones, DACs & AMPs'.

 ## CLEANING UP RECORDED AUDIO

NO MATTER, WHAT FILE FORMAT YOU USE, it will always be true that recording needs to be made in an acoustically treated space, with good gear, and then it needs to be cleaned up/de-noised. If a good job is done of cleaning up the audio, a down-sampled lossless file (with a bit of dithering applied) down to 16-bit/44.1kHz will sound pretty darned good. The dithering adds a little noise to mask any aliasing noise caused by the down-sampling process from 24-bit to 16-bit.

Here is a visual example of cleaning up/de-noising a file in Izotope RX. In Izotope RX, this is a module called Spectral De-noise.

You select some parts of the file where there is no recorded sound, so what you are selecting is just noise. Then using the Izotope's RX Spectral De-noise module, you might opt to select High Quality - Slow, then select Learn. That learns into computer memory what the background noise is made of. I like to knock down the default settings so the Reduction is under 6 dB. Likewise, I reduce most of the advanced settings a little. I always want to go on the lean side - you can make it completely devoid of background sound, but then it will not sound natural:

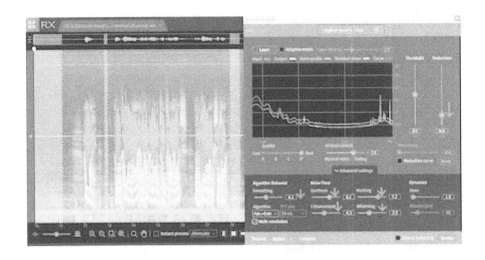

Then I Select All, and Render. This reduces the noise profile from the whole file, as seen.

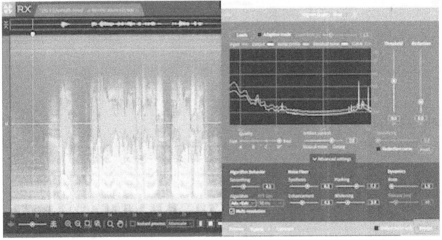

You can see that the resultant file looks a little cleaner, not drastically, but it is significant. I like to make another pass in the more troubling bass area.

I again select noise only sections, this time below about 150 Hz, where bass issues are more dominant. The horizontal lines show these modal or otherwise electrical noise type issues. Again, it is Learned, then I Select All. This time I select one of the menu pre-selects, 'Reduce only Tonal Noise or Hum', then Render. This De-noises the tonal and hum type noise in that noise profile.

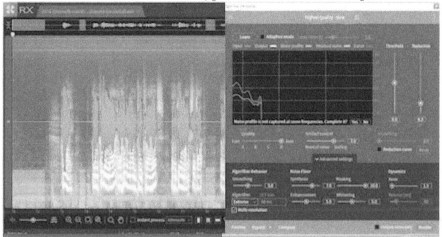

The result, below, is pretty clean. You might notice the haze and reference lines above 30 kHz. This is because the file is 192 kHz. As I understand it, some of the noise reduction algorithms shifts some noise into this above-human-ultrasonic hearing area.

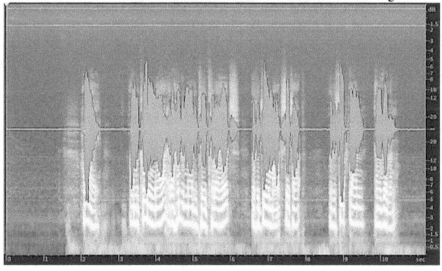

BUT AS EXPECTED THERE ARE STILL SOME FAULTS IN THE FILE, rustling of clothes, paper sound, swallowing, etc. I listen to the file and where I hear any odd sound, I select a possible visual culprit with the rectangle select tool, then Play it, which plays the sound just in the selected rectangle. When I correctly isolate the sound, I again select Learn, and then Render, but while keeping the selected area selected (not Select All). I might apply it twice, until the space looks like the surrounding area/sound. This specific sound I was isolating sounded like a slight tap of a foot against a chair leg:

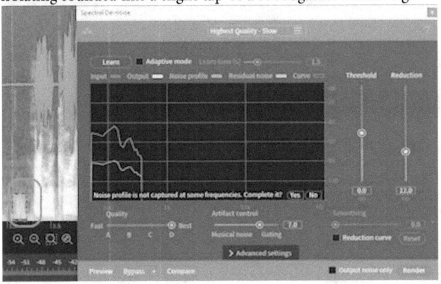

You can see the lighter spot is gone. Also I worked on the narrow vertical line above it, which was some kind of slight click sound. Sometimes the bit you want to remove is too small to Learn (you get an insufficient selection error message) - in those cases I delete it if it is in open space.

I prefer cleaning the file up this way. It is less intrusive than applying specific modules across the whole file - having said that, the Voice De-Noise module is actually very good. Applying a number of module effects across the whole file can accumulate to an unnatural sound, so it is good to be cautious, notching down default settings, applying multiple smaller setting passes, using Spectral De-noise before other RX modules, and not overcleaning it to an unnatural state.

Remembering that this process is permanent (destructive). After doing this process, I can select and post it back into my DAW (or export a file and import that into the DAW. In the DAW, I might then experimentally apply the Izotope modules across the whole track, such as De-plosive, De-ess. In a DAW those effects are non-destructive, meaning I can experiment with the settings, changing my mind over time. As I said above, you can clean it completely of noise, but then it starts to sound unnatural, so you are aiming for subtlety. If you have to make large de-noising actions it will downgrade the overall quality of the audio, the remedy being room acoustics, good recording gear, and subtle de-noising of the file, as above.

 **PLAYBACK**

HERE, I WILL PICK UP AGAIN, ON THE PLAYBACK SIDE, aspects for achieving quality audio:

1. Have suitable and good quality DAC/(pre)AMP and AMP, and speakers, and headphones. Likewise, Class AB musical quality amplification if you have an AVR.
2. The speaker wire, cables and connectors are important, least they become the weakest link, so take some measures to ensure there is no noise on your lines, and warm, neutral or bright according to your gear and preference.
3. Focus on some analog parameters for its emotional connectivity, for example R2R technology in your DAC, some DSD* files, LPs and Phono player, Tube and/or hybrid Tube preAMP and/or Tube AMP, tape. You can simulate those such as with tube-based VST plugins in your media player (aka saturation control). In that case use JRiver Media Player, see https://jriver.com/ or Roon, see https://roon.app/en/, because they accept VST plugins, or Roon. JRiver does not stream and Roon is expensive, or similarly I have suggested the Schiit Loki EQ.
4. For streaming, have a good connection, and access lossless HD and Hi-Res files to the extent you can, which requires higher level streaming providers and upper level plans. Adjust the Buffer Size sampling rate on your DAC to avoid any glitching.
5. Acoustically treat your listening space as much as you are able and decor priorities permit.
6. Use (spinorama) EQ for headphones. See the chapter on Equalization and Home Listening.
9. EQ Room calibration, more so for movies than stereo music,

will be helpful if you listen from the same couch or seat. See more on this in the chapter on 'Equalization and Home Listening'. I do not like Room EQ because it only works for one listening spot, making other listening places even worse.

* **DSD**64 (file extension .DFF .DSS), is the digitizing format for Super Audio CD (SACD), which never took off. Many audiophiles swear by it, so, it's worth a try. Nowadays, the DSD playback standard for DACs commonly goes up to DSD512, which is 512 x 44.1kHz, though you really only need half that, DSD256. It may have been recorded directly in DSD or in edited DXD (352.8 kHz - 8 x 44.1kHz), then resampled to DSD. Often, at the below sites the recording and mastering details are provided, which is helpful, because at this level that information matters in terms of final sound quality. You can buy DSD recorded material from certain websites, such as:
Native DSD: www.nativedsd.com/
Channel Classics: www.channelclassics.com/
Blue Cost Records: bluecoastrecords.com/
Forward Studios: www.forwardstudios.it/en-gb/store/Music-c95566325
Pro Studio Masters: www.prostudiomasters.com/ [All types of formats)
Chasing the Dragon: chasingthedragon.co.uk/shop.html
HDtracks: www.hdtracks.com/
The DSD system is crucially different than PCM (the usual method/standard of digitization). PCM uses a sampling rate on sound waves to achieve a representative code. DSD does not sample in that way. Instead DSD takes a 1-bit sample, using a different kind of technique called pulse density modulation. In doing so, it handles an even higher dynamic range than PCM, thus more detail, thus more nuances of sound, and the method does not create any quantization noise. DSD is more representative of an analog sound. See the chapter on 'Headphones, DACs & Amps', for more on that, and some recommendations. Here are some free DSD downloads:
www.nativedsd.com/free-dsd-download/

ANALOG AND DIGITAL

BEFORE DISCUSSING STREAMING IN THE NEXT CHAPTER, IT IS
WORTHWHILE DISCUSSING PREFERENCES FOR ANALOG AND DIGITAL, in
the first place.

Analog is well known in audio in terms of real-to-real tape
players, vinyl LPs, cassette tapes, and tube amplification, less so in
terms of the DSD file format and certain types of DAC
technology. Analog for many has a distinctive airiness and
presence which, above all, is more viscerally connecting. Nobody
says why, exactly. My own view is that it is about the increased
physicality of those technologies, which matters more when the
music is musically resonant, as in jazz, acoustics or classical
music. In the late 1990s Sony was developing a method of
encoding analog sounds into digital waveforms as an archiving
format, called DSD, since original tape recordings would
deteriorate over time. Then in 1999 it became the stereo digitized
format for Super Audio CD (SACD). SACD never went
mainstream because of industry cost effectiveness along with
forthcoming streaming technologies. However, you can still buy
SACD players, though usually pricey, and you can buy SACDs.
Apart from SACDs, DSD files are very big, so you have to buy
them and play them off local storage.

PCM sampling, the mainstream digitizing method, was
established earlier, is more digitally processed, is multi-channel,
and accurate, perhaps too accurate for some genres. It is used
more because you can apply digital processing effects like EQ, and
so on and so on. Not quite true! Many DSD files start as DXD
(very hi-Resolution 352.8 kHz PCM, are edited, then mastered to
DSD.

Meanwhile PCM has kept up with modern trends with ever
increasing Hi-Res offerings, and ever advancing DAC processing
technology. Hi-Res PCM can be played on technology that
attributes some aspects of analog. So it is hard to say very pure
digital PCM played over, tube amps, for example, lacks anything.
Rather, the benefit is more noticeable in the quality of your room

acoustics and your gear quality.

Meanwhile PCM has kept up with modern trends with ever increasing Hi-Res offerings, and ever advancing DAC processing technology. Hi-Res PCM can be played on technology that attributes some aspects of analog. So it is hard to say very pure digital PCM played over, tube amps, for example, lacks anything. Rather, the benefit is more noticeable as the quality of your room acoustics, and your gear quality creeps into hi-end.

Movies used to be filmed in super 8, which used analog film stock (film strip), with Hollywood generally moving to digital by 2013, which saved time; and time is money, and digital does not deteriorate over time as analog film stock does. Basically, the same story as music. However, some producers, such as Quentin Tarantino (a.k.a. Pulp Fiction, Inglourious Basterds, Once Upon a Time in Hollywood (2019), Steven Spielberg being another big name, continue to use Super 8 cameras, produced by Kodac. Super 8 utilizes newer electronics processing technology, but retains analog film stock. There is something special, something visceral, about analog warmth, in video as in audio. Have a look at Pulp Fiction on Netflix and ask yourself if there is a visual quality that is more connecting than many other movies. There is a kind of visual difference that is hard to put your finger on. And, the argument is that there is a visually connecting aspect - it's not a clear thing, but a subtle nuance worth considering.

As the internet of things grew; and digital technologies matured, various lossy compressed file formats were developed for streaming audio, and video, over the Internet, overcoming bandwidth limitations that existed at the time, and still do to this day (depending on whether your countries has fiber-optic network, or not). Buying these compressed digital files online and downloading them for playback was popularized before the recent burst onto the market of Audio on Demand and Video on Demand services. Meanwhile, the fiber-optic Internet backbone continues to expand so that video and audio is downloaded much faster. Netflix (and others), and Spotify (and others), and other national streaming services have mostly taken over the buy and download market with conscription based click and play streaming services, and cable TV. These services are extended for playback on multiple devices, including tablets and smartphones, and newer smartphones can accept the high 4K and Hi-Res video offerings. Then Tidal came out with the MQA format, which does compress, but in as true-to-native digital quality as they can come up with, therefore, they can transfers hi-res efficiently over the Internet, with compatible devices fully unfolding/uncompressing the file, to the 'so called' full Hi-Res. In part MQA is industry supported because it protects the original file from being copied. The idea is to keep it licensed, and the mastering process, therefore, quality controlled. However, audiophiles have been critical of MQA, indeed Tidal are now (2024) moving back to a focus on Lossless audio, indeed in 2024 streaming lossless hi-res is a viable option.

Meanwhile, old home libraries of CDs and DVDs are being turned into digital format by owners wanting to store their treasures, and that includes SACDs ripped to DSD64 files.

Analog still offers something unique by way of a nuanced more emotionally connected experience. LPs, even with scratches and a slight hissy noise still claims to be more emotionally connecting. LP lovers are still investing in LPs and Phono players, there is even a resurgence in cassette tapes, and many are pursuing Tube amplification in the playback path, perhaps compensating for the modern trend of mastering to a tending toward an overly bright tone to get the best results from mobile playback devices and cheaper mainstream gear. And, there is a market for DSD files as preferred over even uncompressed PCM.

Meanwhile DACs for listening through headphones and DACs inside the ever evolving AVRs and Soundbars are becoming compatible with DSD and MQA. In the case of AVRs and Soundbars, Dolby Atmos has raised the 3D virtualization stakes, though only when video streamed is presently lossless (2024). Atmos can be lossy by way of Dolby True HD, up to 24-bit/192kHz, packaged in Atmos, but that is presently only realized for video though Blu-rays and Ultra HD 4K Blu-rays. For music streaming you can get lossless versions of Atmos from streaming services like Tidal and Apple Music.

While Atmos is a real immersive experience for movies, music lovers often prefer to be in front of the stage, rather than on it, so stereo lossless formats should continue to be a big thing.

Notwithstanding, spatialization and binaural technologies are being implemented and refined in audio production and playback. Apple are probably the key provider to watch on this one with developments of their Apple Spatial Audio (their take on Dolby Atmos).

Analog exists along with hybrid technologies that utilize electronic processing - tubes are built into hybrid playback systems, while some audiophiles stick with fully analog tube amplification, tape-to-tape players, and LPs.

For more serious listening, analog pleads for a place, if not natively, at least simulated and/or in hybrid technologies that blend analog and digital.

SUMMARY

I STARTED WITH a list on the recording and post-recording side, on how to achieve quality audio. Then we began to look at noise from the point of view of analog vs digital, because not all noise is bad. Some noise is wonderful. Then, not just what Hi-Resolution audio is, but how that directly relates to the digital conversion process. Once in, audio needs to be cleaned up, which we looked at through RX Izotope processing. Then to playback. We, again, listed how to achieve quality audio playback, and, related to quality we went above high-res to discuss the, as close as possible, analog native DSD digital format. Analog matters. Keeping things analog, commonly in hybrid form, is important for an emotional connection to music.

CHAPTER 10
STREAMING AND PLAY-THROUGH

"What's been missing from digital music sales has been the possibility of added depth", by David Bryne.

THESE DAYS, much of our audio and video content is delivered to us via streaming. This chapter covers what we need to know about this, from the formats, the services, including some common areas of interest: hi-res audio, Atmos, and multi-casting. To a large degree digital is coming of age where it is more able to maintain that illusive 'added depth' of analog. An initial overview was explained in the chapter on 'Headphones, DAC/AMPS' in a section called 'Getting Hi-Res music from streaming services'. In that case it immediately related to headphones and DACs because of the common purpose of listening to music on those devices through a streaming music provider. In this chapter, I cover the formats involved in streaming, the delivery process, 4K video with

Hi-Res, Home CD library and music streaming services, and Multicasting, whole-home audio systems. These all relate to streaming and play-through.

 **STREAMING**

AS WE KNOW, MANY STREAMING SERVICES can be run on any device, anywhere: smart TVs, computers, smartphones, tablets, and AVRs, and other devices. In terms of music, as of writing, Spotify, and YouTube Music offer a free tier. In that case, you live with the advertisements and reduced services, removing the adds and getting better audio quality by upgrading to premium services. Other providers, these days, often offer limited time free access deals, typically with an opt-out-at-the end of the month option.

Compressed audio is slowly slipping into the past, except for MQA (which is a new compressed hi-res standard, which works well when played on a compatible player to fully unfold it). Compared to the size of streamed video, lossless (non-compressed) audio takes up a lot less space; and, lossless audio is always better than compressed audio.

 DIGITAL FILE FORMATS TO WATCH OUT FOR

IN BASIC TERMS, AND BUILDING ON THE CHAPTER 'CLEAN AUDIO' TO BROADLY COVER FILE FORMATS:

- **Compressed file formats** are not as clear as uncompressed (lossless) formats. A common compressed audio format is **mp3,** and a newer one is the **.OGG** format, originating with Spotify, the .OGG format also has a variable bit rate, so if you are on the Spotify Premium account the audio quality, though still compressed, gets better. For social listening such as in a car, or while on a bus, or for background music, compressed files should be fine.

- **Uncompressed lossless files** are better sounding, always.
- **16-bit/44.1kHz** is CD quality, and so is technically considered as clear as anyone would need, but it is not as good as it could be for better lounge or headphone listening, because of a lack of tone, and comparative amount of noise.
- **24-bit/48kHz** is commonly offered as lossless downloads from music on Demand Services, and is a very good entry hi-res standard.
- **24-bit/96kHz** is in Hi-Res territory and what you are getting variable minimum percentages of on services like Qobus, Tidal, Apple Music. If the recording and mastering process was done well, and you have good DAC which does a good noiseless conversion to analog, you should hear an improvement in terms of clarity and tone. Then, above that format, it gets harder to hear any improvement, but you can if you pay attention to room acoustics, higher level gear.
- **MQA**, which Tidal uses is compressed for high-speed streaming, and the decompression methodology is very good, basically, it contains all the information of the lossless files, temporarily transferring it in compressed format, fully unpacks it in an MQA compatible DAC. The only thing against it is that that breakdown and reconstruction process could add a little noise, not that the average person could distinguish it from lossless audio. Perhaps because of audiophile critique (and increasing capability of Internet transfer), Tidal is now heading back in the direction of lossless audio.
- **Dolby Atmos**, is a different beast, nicely compressed, adding height channels to surround sound for a seriously immersive experience, and reconstructed into a spatial mix on headphones. Even though it won't be Hi-Res when streamed (more likely you'll get16-bit/24kHz; perhaps you'll get 24-bit/24kHz), the difference in being on the stage (immersive) versus being in front of the stage in Hi-res stereo may be a shift that many listeners prefer, especially governing that many listeners are not set up for Hi-Res in the first place. An invitation into a virtual music studio or a virtual orchestra pit is hard to turn down, at least some of the time. Mostly, the idea of Atmos is for enhancing movie special effects, not music, though music is increasingly being released in

the Atmos format. Even though Atmos can incorporate Dolby TrueHD (which is variable Hi-Res audio, up to 24-bit/192kHz) that is not commonly streamed (2024) probably because the multi-channel audio makes the files a lot larger. Without streaming, Ultra HD Blu-ray can include Dolby TrueHD 24-bit/192kHz.

- **DSD256** is worth listening to as the closest digital format to analog, with the caveat that it depends on what it was made from. Some music is recorded directly in DSD – then great, and probably available in DSD256, which is great. The SACD format is DSD64, and were often made from quality master tape recordings – then great. Ripped files off SACD played through a DAC, or from a modern SACD player, or DAC/AMP vs the DAC in an older SACD player – then great. DSD files bought from reputable online companies – then great. Conversely, streamed Hi-Res lossless PCM files made from good source tape masters can sound just as great, subject to complementary playback equipment.

 **MAINSTREAM UP-TAKE & FILE FORMATS**

HERE IS A STARTER ON MAINSTREAM UPTAKE AND FILE FORMATS: The prices I quote are very ball park, for individuals, on high level plans, early 2024: **Amazon Music** ($13/month), **Apple Music** ($10/month*), **Deezer** ($12/month) and **Qobuz** ($18/month). They compete by streaming Hi-Res uncompressed music. **Tidal** ($17/month) uses the MQA quasi-lossless hi-res format, requiring a DAC that can fully unfold it.

  * Apple Music seems the best deal if you have an Apple Device, albeit it is a competitive market so do trial various providers.

  Spotify Premium ($11/month) uses the compressed .OGG format at the moment, which is pretty good, and Spotify is free if you don't mind the ads and lower spec audio.

  Video/movie/TV providers may offer uncompressed Hi-Res audio, and increasingly it comes with Dolby Atmos, but, that Atmos is probably 16-bit/24kHz, possibly 24-bits/24kHz.

THEN THERE ARE SOFTWARE PLAYERS on your device: **Plex, Roon, Audirvana,** etc, are pay per month media player containers (except, try Amarra Luxe). They manage Music on Demand services and any local music computer or external drive, auto-displayed for you, playing bitperfect, and seem to add extra clarity of audio, especially through upsampling. Albeit it is a competitive market, so compare with the streaming services native apps.

Roon and **Amarra Luxe** enable EQ – more on that in the 'Equalization and Home Listening' chapter, because there are alternative ways to apply EQ. JRiver is a pay per version similar player, but does not incorporate streaming services, but does allow for EQ. JRiver as an alternative to streaming services, offers a free shareable pool of music on Cloudplay.

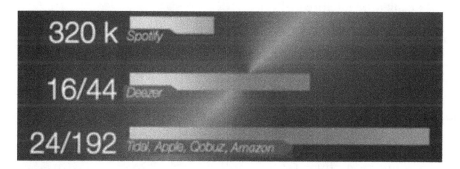

TRADITIONAL MAINSTREAM AUDIO includes:
- Content played on devices like TVs, VCR, CD and DVD players, AM/FM radio, iPod, Stereo receiver/amplifier, AVR [A or A/B amplifier class] providing:
- Surround sound.
- Content via satellite dish.
- DAC/AMPs for headphones.
- Game consoles.
- Compressed audio and video formats for streamed content.
- Early adoption of spatial sound options.
NEWER TECHNOLOGY BEING TAKEN UP INCLUDE:
- Content played on Smart TVs, and/or Apps on any device.
- International and national optical cable installation, enabling much

better Internet transfer capabilities.
- Streamed digital content via modem/router.
- AVRs with Atmos [A or A/B amplifier class, or the more compact class D coming up to a quality standard].
- Improvement of audio and video compression and streaming codecs, including .OGG and MQA.
- Upgrading of game consoles, to include Atmos audio, and virtual reality.
- Dolby Atmos becoming mainstream on top of spatial sound.
- moving to uncompressed Hi-Res audio, and 4K video, 8K video to come. 4K also means you can use the TV as an extended computer monitor.
- Compact devices integrating Wi-fi and Bluetooth connectivity, thus providing more portability options.

TRADITIONAL CONNECTIONS INCLUDE:
- Coax cable (for satellite dishes).
- Optical cables (for surround sound), then HDMI and RCA connections (between players and TV).
- Adapter technology such as to turn a TV into a Smart TV (Chromecast, Roku, smart TV boxes).
- Early adoption of room correction (EQ) e.g., miniDSP.
- Remote controls.
NEWER CONNECTIONS INCLUDE:
- HDMI ARC and HDMI eARC (enabling Atmos and remote control inter-connection).
- older connection options usually still available.
- more powered speakers available.
- Automation of room correction (EQ).
- Increasing control through Apps.
- Digital to analog conversion (DAC) is more sophisticated, improving de-noising of high-res audio in the conversion process.

TRADITIONAL FILE FORMATS INCLUDE:

.MP3 (Lossy): Popular because of its small size, 10 times smaller than CD, becoming an increasingly outdated standard on download stores, convenient for playing on SmartPhones.
.ACC (Lossy): An alternative to MP3, a bit better in quality, up to 24-bits 96 kHz, used by **Apple Music** for their basic offering (at 256kbps) and **YouTube Music and YouTube Movies,** also **Tidal Premiun** (at

320kbps).

.AC-3 (not hi-res): Also called Dolby Digital and Dolby AC-3), Similar to ACC but is a 5.1 channel format for surround sound.

[Note: Dolby Digital was traditionally passed through Optical cable]

.... *Quality improvements*

.Enhanced AC-3, called **Dolby Digital Plus**, DDP, DD+, E-AC-3, EC3, used by **Netflix, Disney +, PrimeVideo, and Hulu,** is up to 7.1 channels, Hi-Res, being *24-bit/192 kHz at 18000 kbps, though more often a practical 16-bit/24kHz, possibly 24-bit.24kHz.*

*[Note: Dolby Digital Plus was passed through HDMI cable. ATMOS became hi-res when it contained D***olby Digital Plus.**

.WAV is an ultimate audiophiles file format, softer at high volume and **.AIFF.** WAV files are mostly for Windows (16-bit/44.1kHz) and AIFF files, mostly for Macs. WAV and AIFF have traditionally been the lossless file formats at CD level (using PCM as the codec).

.ALAC (Cd to Hi-Res): Apple's alternative to FLAC. Used by **Apple Music** (ranging from CD to Hi-Res 24-bit/192 kHz).

[Note: MPEG-4 audio files with an MP4 file extension usually contain digital audio stream encoded with AAC or ALAC. This may not sound as good as a WAV because the processing to unpack the ALAC can still add noise]

.FLAC (ranging from CD to Hi-Res), up to 32-bit, 192kHz, and stores meta data, and takes up half the space of a WAV file. Not supported by Apple. **Deezer** streams FLAC up to 44.1kHz (CD-quality), **Qobuz** and **Amazon Music** streams FLAC up to 192 kHz.

.... *for passing audio through wire*

PCM is an uncompressed 'raw' two channel signal for passing audio through wire, such as chosen as an output setting when connecting a Blu-ray player or disc player to another device. Alternatively if you select to Bitstream a file, whatever it is, is sent without decoding and the receiving player does the coding into PCM or other format. Bitstream is a must for Dolby Atmos files, which are then decoded by an Atmos renderer on the playing device.

NEWER FILE FORMATS INCLUDE, ARE INCREASINGLY LOSSLESS (UNCOMPRESSED) INCLUDING:

.OGG (Lossy): Established by **Spotify,** an audio compression approach called Vorbis, which can also contain video, being a lossy, open-source alternative to AAC. It starts playing before the full file has streamed, and has a variable bit-rate.

.MQA (Lossy and Lossless combined) Master Quality Authenticated, meaning it is recorded in a quality controlled environment (at 44.1 kHz, or 48 kHz or 96kHz depending on the source). **Tidal Masters** uses MQA. Technically, if a source is 24-bit/192 kHz, the MQA method uncompresses it like origami. It down-samples it to 24-bit 96 kHz using noise shaping methods to minimize destructive noise, and again to 16-bit so as to preserve something like 20-bit resolution, and again into two audio streams of 0-24kHz and 24-48kH, and again applying lossy compression, the final result being a 24-bit 48 kHz PCM file with a reversible lossless digitally embedded watermark. Thus a compatible MQA playback device can use the watermark to reconstitute the file as 24-bit 192 kHz, or very close to it. Perhaps the sound is clearer than PCM, but having a less coherent overall musical pulse, and less depth of soundstage, see this critique by Andrew Harrison, arstechnica.com/gadgets/2017/05/mqa-explained-everything-you-need-to-know-about-high-res-audio.

.DOLBY TrueHD, lossless Hi-Res. *Up to 24-bit/192 kHz at 9216 kbps. To get this higher quality you will need Blu-ray.*

[Note: If you are passing .ATMOS, you need HDMI ARC connections]

.ATMOS, includes an .atmosIR file with all the data needed for rendering audio. **Netflix, Disney +, PrimeVideo, Hulu,** playing over surround sound, subwoofers, height speakers; whatever is available up to 7.1.4. 7.1.4 is 7 speakers, 1 subwoofer and 4 height speakers. The 7 speakers are Left, Center, Right, Side-left, Side-right, Rear-left, Rear-right. Modern Soundbars often built to produce 3.1.2., which is Left, Center, Right, 1 subwoofer, and two height speakers. The Atmos renderer will downsample that at adapt to a smaller range of speakers.

.DSD (Hi-Res+), format actually .DSF and .DFF file extension, but is so often referred to as DSD. It is touted as more analog-like than PCM, especially native DSD (even though it is digital). Because of the huge files sizes, it is impractical for streaming, so you buy the files, or SACD disks, and play them locally. See **www.nativedsd.com,** where most of the offerings are classic, acoustic, and solo artist music. Information on the production process is usually included on reputable sites. DSD256 should be as high as you need, native DSD the purist of the two.

STREAMING FILE DELIVERY PROCESS (INCLUDING ATMOS)

AUDIO AND VIDEO PRODUCERS UPLOAD VIDEO AND MUSIC, by account, to On Demand providers. The process by the streaming service companies requires encoding the audio, packaging it up with metadata instructions into a bit-stream container file, like MP4, then streaming it according to a Real-time Transport Protocol.

As an example, in the case of Dolby Atmos an audio editor sends **a master file to an On Demand provider** – called Dolby Atmos Master ADM (Audio Definition Model). That is the specialized container file which the On Demand provider can work with. The On Demand provider then stores it, encodes it with audio codecs, like: AAC, Dolby Digital, Dolby Digital Plus; Video codecs, like: H.264 or VP. The On Demand provider then **packages the necessary file/s into Bitstream container files** like MP4, FLV, WebM, ASF or ISMA – inside the file there can be video, audio, and .atmosIR metadata instructions for your device to unpack. They create multiple output files, so they can feed the right files according to a user's account access level (or click choice if they are buying it). The Bitstream container file, like MP4, is delivered in packets using a **transport protocol,** such as Adobe's Real-time Transport Protocol (RTP), or Apple's HTTP Live Streaming (HLS). The receiving device uses an Atmos renderer to unpack it and pass it out to your designated speakers.

IF YOU ARE DEVELOPING AUDIO CONTENT, and you are interested in Atmos, the question arises as to **how to test** master files before sending them off to an On Demand provider. Not just that, how do you make Atmos files in the first place?

Ref. for info below, by Mike Thornton, at Production Expert, see www.pro-tools-expert.com/production-expert-1/2020/5/30/dolby-atmos-what-hardware-and-software-do-i-need.

EARLY IN THE EDITING PROCESS, you can add the **Dolby Atmos**

Music Panner plugin to your DAW for positioning audio objects in a Dolby Atmos audio mix (thus, creating Bed and Object and panning metadata), see professional.dolby.com/product/dolby-atmos-content-creation/dolby-atmos-music-panner. Processing the audio flow in the DAW is done by having the **Dolby Atmos Renderer** software set up – which can fold the necessary data and files into an MP4 container file, while also creating a Dolby Atmos Master ADM for uploading to On Demand providers. Sometimes you do not need to set up the Renderer because it is integrated into a DAW, presently: Apple Logic Pro, Blackmagic Design's DaVinci Resolve, and Steinberg Nuendo.

To test your MP4 output, Dolby provide a software based **Dolby Reference Player,** for the playback of Dolby Digital®, Dolby Digital Plus™, Dolby TrueHD, Dolby AC-4, and Dolby Atmos®, explained here by Dolby Professional files, see professional.dolby.com/product/media-processing-and-delivery/drp—dolby-reference-player. It can unpack and play DD+JOC MP4 files that are exported from the Dolby Renderer. This would be a good approach to hear how the final output sounds, if you are mixing on headphones.

4K VIDEO WITH HI-RES AUDIO

CAN YOU GET NETFLIX HI-RES AUDIO? More and more content on Netflix is coming out in 4K video and Hi-Res audio, which you get on the higher level account.

If you are playing Netflix on a Smart TV or TV with Chromecast HD with Google TV, you can get 4K Ultra HD which includes Dolby True HD wrapped up in Dolby Atmos (up to 7.1.4). So, when it is available, you could theoretically get Hi-res 24-bit/192 kHz at 9216 kbps; in practice, because of streaming limitations you will probably get 16-bit/24kHz, possibly 24-bit/48kHz.

To get it, you need a plan that supports Atmos. Check in Netflix Account Settings > Plan Details > Premium Ultra HD.

If you are concerned whether your download speed will support it, you can test your download speed at fast.com. Divide the result it gives you by the number of household users that can be on the Internet at the same time. You need 25Mbps or higher per user to stream 4K Ultra HD. If you are on an optical network it should be fine.

To check your TV settings are enabled, as long as your TV supports Dolby Digital Plus and Dolby Atmos, set Streaming Quality to Auto (which will take it to its highest capability setting).

 ## HOME CD LIBRARY AND MUSIC STREAMING SERVICES

AT SOME POINT YOU MAY BE WONDERING ABOUT YOUR MUSIC COLLECTION, which may be on CDs. You may have digitized them, or may wish to do so at some point, perhaps to avoid scratch-aging, or your DVD player is starting to die, so it seems like time to store your ripped music in one folder, it will be easy to play it from your computer, especially if you use a player like JRiver, Plex, Roon, Audirvana, or Amarra Luxe.

The other option is to make a list of all your Albums in a text file, then with a service that plays Hi-Res (Apple Music, Amazon Music, Tidal) you can search for the Albums one at a time, and add them to a Playlist, such as named, "Home Cds". Effectively you've upgraded your home CD collection to remastered Hi-Res quality. But if you cancel that service, you may loose it and have to redo it again with another service – that is not so hard if you keep a text file list of the albums.

 ## MULTICASTING, WHOLE-HOME AUDIO SYSTEMS

YAMAHA AVRs HAVE MULTI CAST, AND WITH DENON OR MARANTZ AVRs YOU GET HEOS. These are whole-home-zone-selectable delivery systems enabled with the backing of your router and WiFi/ethernet connections.

See more information on MultiCast, by Yamaha Corporation of America, youtu.be/PPLrHbVEbC, and for HEOS, see this by HiDEF LIFESTYLE, youtu.be/8Pd5KXWr7Hw.

The Apps themselves make things very easy for simple playback control from a smartphone. For example, if you have HEOS or MultiCast on your AVR, and you just have your audio setup in your lounge, you can still easily access your Amazon Music service, and other streaming services (Apple over AirPlay), using the HEOS or MultiCast App on a computer, phone, or tablet.

If you you use Multicast and upgrade to **wireless speakers,** check compatibility with either MusicCast or HEOS according to which system you have. For example, if you add height speakers to your HiFi system, and want multi-casting, it would be smart to get wireless MusicCast or HEOS compatible speakers.

A CHEAPER SYSTEM, instead of an AVR, is a **Soundbar.** Sonos soundbar systems (sonos.com) also have their proprietary whole-home audio system, see www.sonos.com/en-nz/how-sonos-works. Here is the controlling App, www.sonos.com/en-nz/controller-app. It is similar to the HEOS and MultiCast Apps which also can integrate other streaming services (Apple over AirPlay). And, here is how to add your own music library to Sonos (support.sonos.com/en-us/article/add-your-music-library-to-sonos). Sonos, is Dolby Atmos enabled. The compromise of Soundbars, comes with its small speaker cabinet size, but for situations where simple, cheaper, and compact is your preference, it could be the bees knees.

SUMMARY

WE BEGAN by claiming that uncompressed audio is becoming the norm for streaming, and, for these developments, it helps to know what file formats to watch out for. I build a table to show how newer technologies are developing along with various lossless audio file formats, and what different services deliver (at the time of writing). I delved into how Atmos files are packaged and delivered, not just for those producing Atmos but so we can have an under-the-hood understand of how Atmos distributes itself to your system. We can now move on to considering lounge listening.

CHAPTER 11
LOUNGE LISTENING

"There's just an incredibly rich and interesting relationship between our listening to music and the way our minds engage", by Tod Machover.

WE PICK UP FROM DISCUSSION in the chapter on 'Room Acoustic Applications', extending our knowledge with an understanding of the effects of sound projecting out of speakers, some acoustic aspects of speaker design and speaker placement, even the design of ortho-acoustic speakers as an upward firing speaker solution.

As we step through this chapter, I weave in various solutions: treatment to overcome speaker wall reflections, dialog enhancement, setup scenarios, and questions related to surround sound and Atmos. There is a system going on here, so we need to understand the parts and how they fit together.

CRITICAL DISTANCE

IN A LOUNGE LISTENING SPACE it is useful to consider the critical distance where audio sounds the best. This is where room resonance meets direct sound – a "sweet spot", a certain distance from the main speakers. Inside this sweet spot is the "near field", within which the sound is clear without room resonance or atmosphere; At the sweet spot there is a nice balance between direct sound and room resonance, that balance meaning the sound has a more realistic sense of spaciousness to it. Beyond the sweet spot, room reflections begin to dominate, drowning out audio clarity. Here is a test case, by Geoff Martin, see youtu.be/6Q0joik6E74?t=748. This is a first understanding.

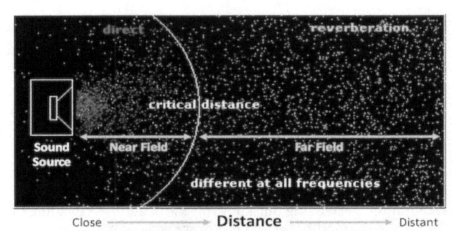

REF: Image from https://hearinghealthmatters.org/waynesworld/2015/visual-guide-to-critical-distance-for-sound

SOUND BEING AT ITS BEST at the sweet spot is also dependent on the size and shape of the room, flat surfaces and acoustic treatment. In the chapter on 'Recording Spaces' RT60 was explained as a measurement for how long it takes for sound to fall away, with an RT60 of about 0.2 seconds being a common preference for a typical recording space in an apartment or house. That is quick fall away time is a preference for recording, but not for listening. That 0.2 seconds quick fall away time is a preference for recording, but not for listening.

For media listening, usually in a lounge living room you want the fall away time to be below an RT60 of about 0.5 seconds. Most domestic living spaces, with carpet, curtains, furniture, and people will average at about 0.4, so that is good. If it is longer than that the sweet spot is going to be crammed up too close to the speakers to be a realistic listening position. And, it is best to have some acoustic panels at direct reflection points, and if possible some bass trapping to reduce bass room mode build up. This will ensure your home listening environment is enjoyable. In practical terms people often sit in different places in a room, not in an exact listening position, so, again, finding ways to add some acoustic treatment will really help that realistic compromise.

 SPEAKER DISTANCE FROM WALL

LOW FREQUENCY SOUND EMANATES in all direction, and actually curves behind the speaker as well as forward, thus the forward direct sound of it to your ears, and the reflection from the wall behind the speakers will be out of time at the listener position. That is problem one. Placing some acoustic treatment on the wall behind your speaker minimizes that. And it minimizes problem two. Problem two is boundary interference where, depending on the distance from the wall, a particular frequency's 1/4 wavelength reflection meets its same frequency direct sound at the front of the speaker, canceling itself out. Here is a video explaining the second issue of boundary interference, by GIK Acoustics (youtu.be/T10_MLGOBfc):

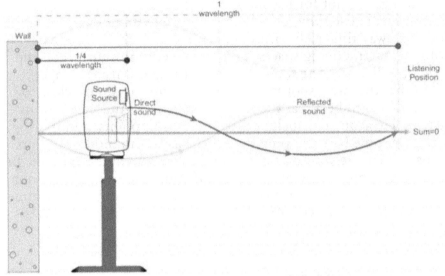

REF: (www.genelec.com/monitor-placement)

SPEAKER MANUFACTURERS, for standard speaker designs, suggest moving speakers either a long way from the back wall (>2.2m) or within one meter (<1m), but if you get too close to a flat wall the low and mid base tend to increase by 2dB, thus fixing one problem, and causing another.

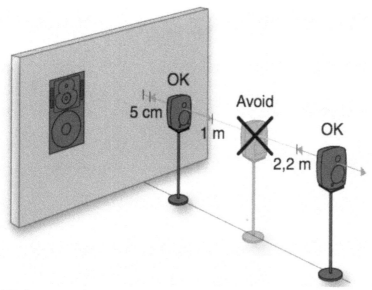

REF: (www.genelec.com/monitor-placement)

Closer, about 0.5m from the wall, pushing the boundary interference cancellation issue to a higher frequency, because then acoustic treatment is more effective. The solution is really the same as above, to put acoustic material behind the speaker, and here we will come to understand how thick it should be. Let's say your speaker's face is 50 cm from the wall, as above, the 1/4 wavelength frequency would be 170 Hz. So, we can put a 180mm thick absorption panel, made with glass wool wall slab home insulation (~ 20kg/m3; perhaps R2.2 from you home depot). It would be a 70% effective solution as seen by entering in these details at acousticmodelling.com:

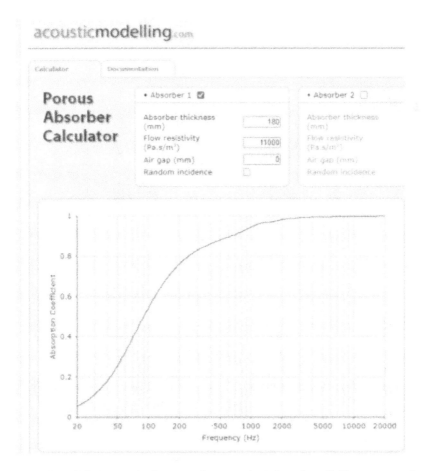

It is tricky to get the gas flow resistivity for different products for entering them at acousticmodelling.com. If interested for different products, see this pool of gas flow resistivity data on the Gearspace forum (gearspace.com/board/studio-building-acoustics/625978-common-gas-flow-resistivity-numbers.html).

Here is a DIY option to do the trick, by Acoustic Insider, see www.acousticsinsider.com/blog/best-insulation-material-diy-acoustic-absorbers. Get R2.2 acoustic wall slabs, build a frame to fit the size of the slabs, adding breathable material to the front. Hemp/Hessian is cheap and works, but acrylic material is also good, as long as you can blow through it:

For the above discussion, I used:
Sound Wavelength Calculator, see
www.omnicalculator.com/physics/sound-wavelength
And the Porus Absorber Calculator at Acousticmodelling.com, see
www.acousticmodelling.com/8layers/porous.php

 +2 DB BASS BUMP CAUSED BY FLAT WALL

HERE IS A SOLUTION TO COMPENSATE FOR THE BASS BOOST that
happens when you are close to a wall. I have said that for any
bassy sound, let's say 80Hz - 250Hz, that can partly curve back
and reflect off the wall, any rebounding off a flat wall will increase
the volume of those frequencies by about 2 dB. Some speakers
provide a setting to allow for this. Here, on the back of my
MR524 there is a switch to adjust the mid bass down by 2dB, plus
a few other settings depending on how your speakers are placed.

But, we realize that if you have acoustic material behind your speakers, this is not necessary:

 WAVE WIDTH AND TWEETER ISSUES

FREQUENCIES ABOVE 3,000 HZ have narrow wave widths, so tweeters which produce them need to be aimed at the listeners. For standard speakers with direct projection, to hear both the tweeters, you need to be listening within the tweeter's projected convergence:

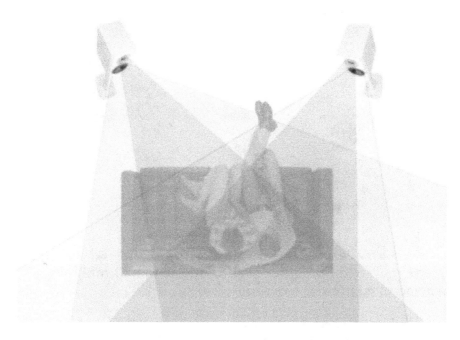

The darker color represents the tweeter's projection, the lighter color, the mid range drivers.

If you are sitting off to the side, you will miss out on the tweeter detail. This explains why some toeing in of typically designed main speakers towards a central listening position is a good idea.

 ## SPEAKER DESIGNS

THERE ARE ALSO ASPECTS OF SPEAKER DESIGN that aim to compensate for narrow tweeter projection, and more. Here are five important solutions, all being beneficial, to improve the dynamics of audio directionality, and clarity:

i) acoustically treat your room.

ii) wear headphones, so the next issues don't matter.

iii) have a tweeter waveguide (rounded out tweeter cones, and the seat the cones are in), thereby widening the higher frequency wave projection:

Ref: The Jantzen Audio Waveguide for Audax TW034 tweeter

iv) have multiple speakers, as in surround sound, or multiple angle mounted tweeters (see ortho-acoustic, below).

v) have a hybrid tweeter lens design which, as well as the wave guide, further disperses the projection. Here is a patent from as early as 1973, on this aspect of design, an acoustic high frequency lens, by inventor Edward M. Long, who is now inducted into the TEC Awards Technology Hall of Fame, see patentimages.storage.googleapis.com/48/9c/4e/8fd9c15402df34/US3735336.pdf.

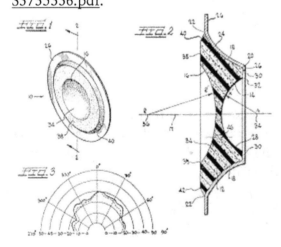

Have a look at the patented DTX acoustic lens in the tweeters of the active SA legend 7.2 Silverback, from Cool Scandinavian Loudspeakers. It takes the high frequency 'narrow' waves, and radiates them through multiple honeycomb paths to a concave lens which disperses the narrow soundwave in a wide array – sort of like the spout of a watering can. As they say, "the sound covers the entire room. You do not have to sit in a certain place to get the best experience." system-audio.com/product/sa-legend-7-silverback/#dropdown.

vi) have ortho-acoustic, upward firing-speakers (discussed below).

This sort of design has multiple tweeters, aimed at different directions, to cover a wider field.

HERE ARE SOME DESIGNS where various of these solutions are implemented:

- **Dutch & Dutch 8c** (www.dutchdutch.com/8c/) While high end, $14K a pair, these speakers use a boundary coupling technology, as per this review, by White Sea Studio, youtu.be/aK-gdXCJoVg, and another by Darko Audio, youtu.be/fygm3iV_odA [active, integrated back firing subwoofers, DSP for EQ, waveguide tweeter, active room matching, side ported].
- **AudioNote AN-J/K/E** (www.audionote.co.uk/loudspeakers) [shallow reverberant cabinets, wide baffle shape, rear ported]
- **Larsen** (www.larsenhifi.com/) [ortho-acoustic, upward tilted, rear

ported]
- **Ohm Walsh** (www.ohmspeaker.com) [omnidirectional, bottom ported]
- **Valutronic SQ50c**, *DIY kitset*
(www.valutronic.com/E/sq50c.html)
 [ortho-acoustic, upward facing, multiple tweeters facing different directions, rear ported]
- **Pi Speakers**, *DIY kitsets* (www.pispeakers.com). [horn tweeters, front ported, suggests to use with a valve amp].

SIZE OF SPEAKERS

BIGGER IS NOT ALWAYS BETTER, AND SMALL CAN BE TOO SMALL. If your speakers are too small for your room, the dynamics of what they produce will be underrated, meaning the highs and lows of sound will not be expressive. If you get speakers that are too big for your room, they will put out too much bass pressure, overloading your room with boominess. This is nicely explained by British Audiophile, speaking about HiFi Myths & Misconceptions - Speaker Setup, see youtu.be/VHOYXjVJKKY?t=590.

What you want is a speaker that will play as loud as you will ever want to hear, as far away as you will ever be, without excessive distortion, in the particular space you have.

The best way to know the right size of speaker for your room, is when you are interested in a certain brand of speakers, to email the company and ask for their size suggestion according to your room size and listening interest.

BRINGING DIALOG FORWARD

FOR AN AV RECEIVER, it is normal to have a center speaker, which is placed just below your TV screen so no matter where you sit in the room the dialog seems connected to the speaker. On the AVR you can adjust the volume of that center speaker in relation to the other speakers.

Your AVR remote may also have a "Dialogue enhancer". For personal preference that makes it easy to adjust dialog volume in relation to other sounds. Sometimes on movies you want to temporarily do that because they are often edited for movie threatres where by for a long the bass can be too loud when there are explosions when people are also talking, for example.

The same applies for music where you want to hear the words more clearly, perhaps if your room is overly reverberant, perhaps just right for classical music, but not for dialog, you can enhance the frequencies where human dialog is clearer, thus effectively pushing out the sweet spot for dialog rich content.

Upper level Denon and Marantz AV Receiver models, feature a dedicated "Dialog Enhancer" or called "Dialog Level Adjust" setting, which will boost the vocal frequencies in the range of 1kHz-3kHz:

Another way to enhance dialog is to apply dialog enhancing EQ by way of a VST plugin. I have used JRiver Media Player through which I have routed audio from whatever audio I play on my computer. JRiver accepts VST plugins, so does Roon and Audirvana. Therefore, if I want to bring dialog forward, I have used the McDSP SA-2 VST plugin, see www.mcdsp.com/plugin-index/sa-2. The McDSP SA-2 has been used to improve dialog in the production process of many movies. It is equally useful for vocals, drums, and any musical application where you want to pull back on the resonance. This works a treat. To secure the registration for this specific VST, you will also need the free iLok License Manager software, see www.ilok.com/#!license-manager:

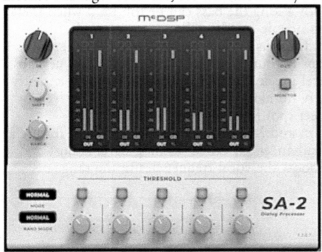

SOUND STAGE AND IMAGING

ACHIEVING A GOOD SOUND STAGE AND IMAGING is one of the goals of hi-fi audio clarity. If you listen at the position that has both speakers at an equal distance from you, as an equilateral triangle, the dialog and music will also seem to project from the center of the two speakers. This is often referred to as the 'phantom center'. It should sound as if there is a stage behind the speakers, or at the speaker wall, generally. It seems as if your speakers disappear, with the sound emanating as a "sound stage".

When you have a sound stage, **Imaging** is then possible. This is how well from within a sound stage you can pin point the position of a particular sound (eg vocals, guitar, keyboard, different drums, and cymbals). If you have created a sound stage then imaging should exist - it also being enhanced by good DAC and amplification technology.

It is often the case that the original recording was done with multiple instruments around a room, and/or in post production, reverb, echo and panning were applied for imaging purposes. Our ears can hear those effects, thus locate the sources of different sounds. Also, for surround sound, editors allocate certain sounds to the left or right speaker channels, surround sound channels, and Dolby Atmos channels. If you have an AVR or Soundbar, that channel information is rendered to the different speakers for an immersive experience:

REF: https://www.masteringthemix.com/blogs/learn/how-to-balance-all-the-elements-in-a-mix

SUBWOOFERS

SUBWOOFERS PROVIDE THE FINAL DEEP BASS bass that your standard tweeter and mid range woofer cannot do. These cover sound down to about 80 Hz. Bigger speaker cabinets with better quality midrange woofers may be quite efficient down to 40 Hz, but to get down into very low bass you are going to need a subwoofer.

LET'S SEGUE TO AN ALTERNATIVE, TOWER SPEAKERS:

A slightly compromised option for movies, but no compromise at all for music are tower speakers. They have larger cabinets and extra woofers (usually a little smaller than subwoofers) tuned for bass, providing a good all round integrated solution, especially for music. Looking at a quality brand of tower speakers, at a reasonable price, and good reviews, see the SVS Prime Pinnacle, https://soundgroup.co.nz/collections/svs/products/svs-prime-pinnacle-speaker:

They go down to 29Hz, which is fine for bassy music, and for movies actually, and you can connect them up to a reasonable level 8 Ohm AMP or AVR. Yes, you could add a subwoofer if you really want to, to get body thumping bass movie effects, but for many people this compromise will provide an excellent option.

LET'S SEGUE BACK TO SUBWOOFERS:

It is far more common to add separate subwoofers to your system, whether for cost, practical or preference reasons. You might get one, then add more as your budget allows. Certainly, two are always recommended. They allow for placement variation, even better bass management, and this approach is perhaps less obtrusive than tower speakers, for you. More importantly you will be able to EQ separate subwoofers in a number of ways (EQing subwoofers is a great idea).

Subwoofers are specifically designed to be efficient in the range between about 30 Hz (or lower) and 100 Hz, commonly having woofers 8" or larger. They are particularly important, for bass music, when matched with a hi-fi system, and for movie sound effects like explosions and rumbling.

I have a RELs T/7i subwoofer (predecessor to RELs T/7x, see: rel.net/shop/powered-subwoofers/serie-tx/t-7x/

The RELs T/7x subwoofer is a closed box design for tighter sound [8" (200mm) woofer, 10" passive driver underneath, class A/B amp], compared to ported subwoofers, which are less tight, but louder. Ported subwoofers would be a preference if you really want to shake the room. Have a look at the SVS PB-1000 Pro as a good example of a ported subwoofer [12" (305mm) woofer, class

D amp], see https://www.svsound.com/products/pb-1000-pro-subwoofer:

80 Hz is usually a suggested crossover point, below which a subwoofer is set to come into effect, 100Hz if your main speakers are bookshelf size.

Directionality of subwoofers is not as important as main speakers because very low bass sound waves emanate more in a circle than being focused forward.

It is usual to place separate subwoofers to the side of main speakers. Also, near room corners, or side walls because many low bass room modes frequencies already gather there. If you have two or more they should generally be in symmetrical positions. To balance frequency output, it is worthwhile to EQ subwoofers.

You can prove this to yourself by putting the subwoofer on a chair at your normal seated ear height, then crawl to where you think you could put it, I am suggesting the corner as the best place, see Paul McGowan's advice representing PS Audio, youtu.be/xIEmZA_ruIg. If corner placement is impractical, look for other symmetrical placement, opposite sides of a room, or each side of main left and right speakers. A starting position for a second speaker can be diagonally opposite the first one, but measurement is advised.

Once a subwoofer is in place, the sound frequencies it produces can be measured and EQ adjustments applied.

EQing a subwoofer will provide clear distinct low bass notes below 80 Hz, rather than a boomy one note effect. To learn more about EQ, see the chapter on, 'Equalization and Home Listening'. Adding **a second subwoofer**, or more, helps to fill in any low frequency room mode gaps (nulls). EQ does not really work on room nulls because the dips are too big, but, a second subwoofer's frequency curves will be different from the first (depending on placement). Those differences even out the sound. The physics of that is "constructive interference" (inter-meshing of the sound waves) so, like waves coming from opposite directions merging through each other, but also those waves reflecting off walls provides additional constructive interference. Here is a video explaining the improvement of having two subwoofers and their placement, by SVS, see youtu.be/OXVpy5jaNsk.

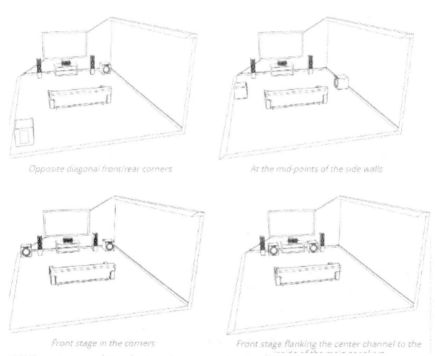

Opposite diagonal front/rear corners

At the mid-points of the side walls

Front stage in the corners

Front stage flanking the center channel to the

REF: svsound.com/blogs/subwoofer-setup-and-tuning/75040195-why-go-dual

So, with two subwoofers, placed in opposite corners, or symmetrically, all the frequencies should be covered.

This following process for aligning subs will not be necessary if you have an AVR that can use Dirac Live + Bass Control (DLBC), which I explain as a preferred system in the next section. If you do not have that DLBC system, it is more technical to manage. You can check your multi-subs work well together, seeing if you need to add delay or switch polarity on one of the subs, measuring with the free REW software, see www.roomeqwizard.com/. This will also help accuracy for, what I suggest should be your next step of, EQing your multi-subs. Here is a YouTube video, by Jeff Mery, showing how to align multi-subs see youtu.be/8bwpLfbLiZ4. This process can include adding a 2 x 4 miniDSP between your amplifier and your subs, whereby you can set EQ and if necessary add some delay to one of the subwoofers.

For more EQ specifics, there is a section on EQing subwoofers in the chapter, 'Equalization and Home Listening'.

UPGRADING TO AN AVR

IF YOU ARE UPGRADING TO AN ATMOS SYSTEM, there are a few considerations. In terms of those amplifiers you may have invested in – let's say they are quality A/B or A type amplifiers – you can still use them by connecting them to your AVR, if it is high enough spec, by connecting your external amps to "Pre out" connections. Two examples with Pre-out settings are the Onkyo RZ50 (class A/B), see www.onkyousa.com/product/tx-rz50-9-2-channel-thx-certified-av-receiver/ and the Yamaha RX-A6A (quality class D), would be another good example, see usa.yamaha.com/products/audio_visual/av_receivers_amps/rx-a6a/index.html. Pre-out settings on the back of an AVR will usually have a white or aluminum colored backing plate. You would also use the pre-outs if you are connecting up powered speakers:

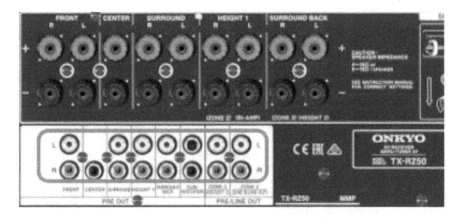

IF YOU ARE GOING FOR A SECOND HAND AVR, with Atmos, you probably won't have pre-outs. In that case, you would have to set aside your existing power amp and rely on the internal amps in the AVR, which you will be doing by connecting your speakers to the terminals with the black backing plate.

Referring again to Dirac Live + Bass Control (also discussed in the chapter, 'Equalization and Home Listening') which is only now (May, 2023) being implemented into a few high spec AVRs, ones that have independent subwoofer connectivity for multi-subs. See the Dirac Live Online store, at www.dirac.com/online-store/, then check each AVR to see which ones can incorporate Dirac Live + Bass Control. If the AVR has independent subwoofer connections, then each subwoofer in a multi-sub setup can be adjusted separately. For example the Integra DRX-8.4, has outputs for four subwoofers, two of which are independent, see integrahometheater.com/product/drx-8-4/, and see this review, youtu.be/Dsn4_Glkxz4.

SURROUND SOUND

SURROUND SOUND, IS MORE OF AN IMPROVEMENT FOR WATCHING MOVIES than it is for listening to music (unless you prefer multi-channel stereo). It enables capturing sound effects inside an audio 3D space. As far as Hi-Fi music is concerned, if it is edited in a surround sound format, this will feel more like you are in a room with musicians, and what you hear will change as you move closer to any one speaker.

Because there are multiple speakers, the far field is partly nullified. It is also common to have a **center channel/speaker** below your TV/display screen, for TV viewing, so that if you are sitting off to one side you will still hear any vocals as if it is coming from the center direction of the screen (rather than a dominant left or right speaker):

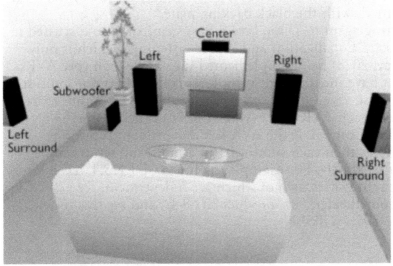

REF: *turbofuture.com/home-theater-audio/How-to-set-up-and-calibrate-your-home-cinema-or-theater-surround-sound-system*

DOLBY ATMOS VS BINAURAL AND SPATIAL

MORE RECENTLY, UPWARD-FIRING SPEAKERS or speakers attached high in the room, are being popularized, as added to a surround sound system to take advantage of **Dolby Atmos.** To get it you will need to add a couple of speakers to a surround sound system, and have Atmos device compatibility.

Even though surround sound, and height speakers (for Dolby Atmos) are not needed for music, this preference might be changing for some, as Atmos for music is coming onto the scene through remastered music for Dolby Atmos. Perhaps some peoples' music listening preference will move from wanting the sense of being in front of a music stage (stereo hi-fi) to being on the stage (surround sound), and then fully immersed with Atmos height channels.

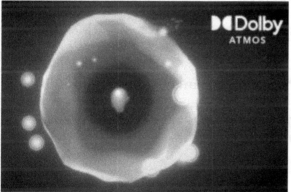

REF: dolby.com/technologies/dolby-atmos/

THERE IS ALSO ADVANCEMENT IN BINAURAL AND SPATIAL RECORDING. I have already introduced the necessary recording technology in the chapter on 'Microphones, Digital Interfaces, and Plugins'.

With the release of iOS 16, Apple rolled out personalized spatial audio, which is for headphones. You point your iPhone's front-facing camera at your ears to capture data on their shape and contours, the end result being a personalized profile through which to filter audio. To take advantage of this you need TWS

(True Wireless Stereo) buds or headphones, like the Apple AirPods. Android 13 has something similar. For wired headsets, go to System settings > Sound & vibration > Spatial audio. For wireless headsets, go to System settings > Connected devices > Gear icon for your wireless device > Spatial audio. On Windows computers you can select Spatial sound settings, with a right mouse click on the audio icon, though that's not personalized.

Keep a look out for new developments in this area.

 TRANSITIONING TO DOLBY ATMOS

TO DEVELOP A SURROUND SYSTEM WITH DOLBY ATMOS, you might:
A, add Atmos for Headphones, for a Windows PC, (Apple have it built in).
B, add or start with a space-saving Atmos compatible soundbar setup for an Atmos enabled TV.
C, add height speakers to a surround sound setup, which may require updating your AVR.

A. Atmos for Headphones (Apple macOS have it built in)

See the section, Dolby Atmos for headphones, in the chapter 'Streaming and Play-through'.

B. Soundbars

MANY PEOPLE, CONCERNED WITH SPACE SAVING AND BUDGET, opt for a soundbar, which goes under a TV screen, containing, in one bar, a number of speakers, likely a left, a right, a center, and perhaps also two upward firing speakers. Added to the soundbar, there is preferably a subwoofer, and even better, additional separate speakers for high-rear mounting.

Atmos compatible soundbars include a renderer to distribute the audio to channels/speakers, and the height objects to height speakers. As a good example, here is the Samsung Q-Series Soundbar HW-Q990B, see www.samsung.com/nz/audio-devices/soundbar/q990b-black-hw-q990b-xy/. With this soundbar, the features include SpaceFit+ which is bass AutoEQ for the sub, and AutoEQ for the other speakers. Look for these kinds of features when looking at Soundbars.

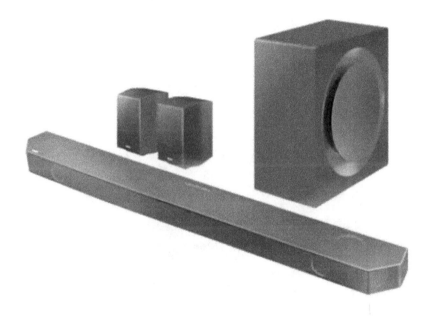

Bluetooth technology with soundbars is okay for movies, because although there is a small delay with Bluetooth, that delay will only be in the effects channels: LFE bass effects to subwoofer, and height speaker channels.

There are a few limitations with soundbars. Firstly, the small size of the speaker cabinets will limit the clarity of audio compared to larger cabinets. Secondly, they all seem to use class D amplification, which is not as pristine a sound as class AB (which needs transformers and heat-sinks) [This is true as of 2023, but class D is improving, so pay attention to reviews of Class D amplification going into 2024 and beyond]. Thirdly, there is not a way to integrate existing speakers or integrate a hi-fi stereo system into a soundbar – for that kind if versatility you need an AVR.

C. Surround sound with an AV Receiver, height speakers added

IF YOU UPGRADE TO AN AVR, you can probably keep any stereo Hi-Fi setup you have, and add a few height speakers. An **Audio Video Receiver (AVR)**, allocates surround sound to speakers via a *renderer*. See the height speakers in this picture:

REF: svsound.com/blogs/speaker-setup-and-tuning/dolby-atmos-and-dts-x-guide

Height speakers may be Blu-tooth to avoid cables. If those height speakers seem intrusive, you can always install in wall or in ceiling speakers.

LOOKING FOR QUALITY HEIGHT SPEAKERS, I suggest checking out www.spinorama.org/, some good options being: Genelec 8030C $500, or for ceiling speakers: JBL Control 24CT $150. I have two Klipsch R-2650-W II In-Wall Speakers, set high in my rear wall. It is best to match brand with your other speakers, though it is not essential for the effects that are produced in Atmos. At least see if you can get something with the same tweeter type.

WHEN ADDING SURROUND SOUND SPEAKERS (not height speakers), it is good to look at quality same brands (or speakers with the same drivers), because the tonal character should be the same when sound is panning from one speaker to another. If, however, your purpose is to make the best of movie effects, thus just adding height speakers, the brand of height speakers and subwoofers won't matter tonally.

UPWARD-FIRING SPEAKERS

THE ORIGINAL DESIGN OF UPWARD POINTED SPEAKERS is by way of ortho-acoustic speakers coming from the famous Swedish sound engineer, Stigg Carlsson (died 1997) under the brand, Sonab. John Larsen, keeps the torch glowing, then in charge of building production; now running Larsen Hifi, see www.larsenhifi.com:

Interest is maintained in Carlsson ortho-acoustic speakers through Carlsson planet, see www.carlssonplanet.com/en/. It is interesting to understand these kinds of speakers, even if you have traditional speakers because there is a relevance to the design considerations of adding "height speakers" for Dolby Atmos, and can be a kind of speaker that works in your situation.

The problems with traditional speakers are, (a) the difference in time between direct sound to the listening position, and reflected sound to the listening position. This timing difference fuzzes the clarity. The issue exists because the speakers are pointed directly at the listener. Although this difference can be solved with acoustic treatment, that is not always practical in a lounge. Then, (b) traditional speakers often need to be away from the wall (depending on design), which is often not practical, or should have acoustic material behind them. Then, (c) because high frequency sound from tweeters has a narrow projection, if you sit other than in the preferred listening position you will miss some of the higher frequencies.

In fact it is worth asking if we get natural sound from the traditional speaker design in the first place. I'll quote from one of Stigg Carlsson's patents, see patents.justia.com/patent/4112256:

> *"The (traditional) loudspeaker shape gives rise, however, to characteristic listening impressions and directional impressions which result in a 'greater directional sharpness' than would be found when listening to natural sound sources in a normal interior listening environment. As a result the reproduction is not experienced as acoustically life-like."*

The Stigg Carlsson ortho-acoustic approach aims/tilts speakers to reflect rather than directly point at a main listening position, with direct reflections reduced by small absorption panels built within the upper-outer cabinet design, or directly behind and/or beside it. Thus the direct reflections are absorbed, and the indirect reflections merge around the same time, creating a wider canopy of audio that is more life-like, without an 'unnatural directional sharpness':

To say this in other words, reading the Stigg Carlsson patent at patents.justia.com/patent/4558762 a key acoustic tonal truth he recounts is that, *"the **reflected sound** which arrives out of phase (time) **with the direct sound from substantially the same direction**, tends to **degrade the tonal quality and the definition** of the sound reproduced. It also tends to **mask the details of the spatial information** provided by good stereophonic recordings ... I have found that a considerably improved reproduction can be obtained, if this elimination of reflected sound can also be extended into the mid frequency range, which carries more important information than the low frequency range."*

Here is a video explaining the Larsen speakers, by Joe N Tell, see www.youtube.com/watch?v=rGpewYJ7plw), and further discussion between John Larsen, Michael Vamos of Audio Skies, Joe Mariano of Joe N Tell, Channa D of Techno Dad, and Michael of Youthman, see youtu.be/mdSjIjjOi5o.

The following Carlsson OA-58 speaker design, by the Stigg Carlsson Trust (SSC), came out in 2004 see, www.carlssonplanet.com/en/speakers/produced/carlsson-oa-58/, followed by an improvised model in 2007 (Carlsson OA-58.2), manufactured by John Larsen, see www.carlssonplanet.com/en/speakers/produced/carlsson-oa-58-2/. New models are manufactured by John Larsen, see the top-of-the-range Larsen 9, www.larsenhifi.com/en/larsen9.htm.

WE CAN SEE IN ORTHO-ACOUSTIC DESIGN, how the acoustic absorption around the drivers prevents direct reflections off the back and nearest side wall, and floor. This means the reflections are generally arriving as a canopy over the listeners, wherever they are in the room. This is good news for a lounge because in a lounge/common family room it is often not realistic to be able to provide extensive acoustic room treatment, and often people will be listening from different positions. Read this review by MMK, see greenaudioreview.com/larsen-4-2-review/.

CONTRASTINGLY, here is a video of Ohm Walsh speakers, by Z Reviews, see www.youtube.com/watch?v=Y_8xUDYBgZQ, summed up as, "the best of the best". In this case, the OHM Walsh speakers, from America, have their own design of turned-upside-down speakers, directed upwards. Here is the company's explanation of the technology, see ohmspeaker.com/technology/. As with the Larsen speakers, because these designs utilize all-room-reflection to create an immersive sound canopy, the need for acoustic diffusion in the room is significantly reduced:

Note that there is still some built-in backside dampening placed inside the cylindrical protector grating to minimize reflections from the closest wall.

I BUILT four DIY ortho-acoustic speakers:

My first two ortho-acoustic hifi speakers, were build using a SQ-50C kitset available from the Swedish company, Valutronic, see www.valutronic.com/E/sq50c.html.

I added pegs of wood on doubled-up hardboard sheets to the inside of the cabinet, whereby I estimated the extra space and made the cabinets that much bigger from the original DIY plans:

I used natural wool for the insulation stapled on with flyscreen:

I used concrete reinforcement grating for a strong protective covering in case one day someone sits on the speakers by mistake. Also, is the tweeter placement:

The supplied Monacor SPH-170 bass-mid range woofer is a 50 W 8 Ohm quality German built speaker, 6.75", frequency range 38 - 5000 Hz, see monacor-webshop.de/sph-170.html. It has a very linear and wideband sound pressure frequency response.

SPH-170 (10.0110)

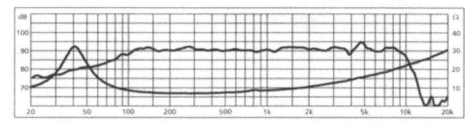

The supplied tweeters, I believe, are silk dome Monacor DT-28N, see www.monacor.com/products/components/speaker-technology/hi-fi-tweeters-/dt-28n/.

DT-28N (10.4020)

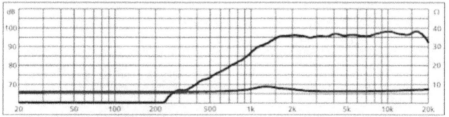

The larger cabinet size provides noticeably better base than my small cabinet design below.

The Monacor SPH-170 woofer is a fairly efficient woofer with about a sensitivity of 90dB, ie can produce decent volume levels without needing a super powerful amplifier. The Monacor DT-28N tweeter has a slightly higher sensitivity of about 93dB, so these are well matched in terms of sensitivity.

I BUILT my second two ortho-acoustic hifi speakers, from another SQ-50-20 kitset available from the same company [same drivers] Swedish company, Valutronic, see www.valutronic.com/E/SQ5020.html.

299

In this case I modified the cabinet design so they would be room-corner-firing speakers, aimed overhead, toed in. I added some internal bracing, similar to the Stig Carlsson and Larsen designs. To make the driver housing I made a cardboard mock up to fit on the top of the speaker cabinet:

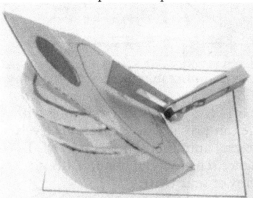

When I pulled apart my cardboard mock up, I could see what wood pieces I needed to cut. I then used a technique called kurfing, to cut splayed groves with my jigsaw into a 16mm piece of MDF, as calculated, see www.blocklayer.com/kerf-spacingeng, and bent and glued it to its shape. Here is a part of that process:

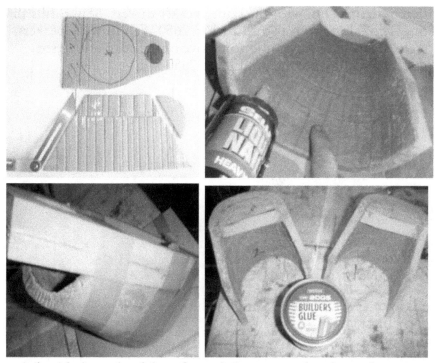

I doubled up the driver face, gluing some aluminum between the two pieces (for rigidity and dampening).

The cover (below), was also achieved by making a cardboard mock up. I used an aluminum track, cut like kurfing, with a hack saw, on the inside lip, and bent it to shape, gluing it to a wooden top-and-backing support, then I inserted some black mesh grating – a bit tricky to do – labor of love:

So, in my lounge I have four ortho-acoustic speakers, of slightly different designs but with the same speaker drivers. Sound fills the room. This, along with two Klipsch R-2650-W II in wall speakers, bought second hand), positioned above our lounge doors, see https://www.klipsch.com/products/r-2650-w-ii:

And, I got a Wharfedale Pacific Evolution center speaker (bought second hand), see:

I tried to do the best I could to match the speakers, with all the tweeters being soft dome, the woofers as close as I could get to the Manocor ones from my DIY building. With a center speaker and my two in-wall speakers, and my RELs T/7i subwoofer, discussed above, I can get Atmos via my AVR, and for stereo music my Denon AVR is set to play stereo through the two main upward firing speakers or mutli-channel stereo if selected. I have had a stereo amp connected, but since my Denon AVR X2200W has A/B class amplifiers it is as musical as the stereo amp I previously used. To up the anti on my stereo sound, I am getting a Schiit

Frey+ preamp, see https://www.schiit.com/products/freya-n.

SUMMARY

WE BEGAN by understanding the listening sweet spot of critical distance, the time it takes for sound to fall a way, thus understanding the physics of room resonance, something to be tamed. Speaker placement was considered, the problems involved and the solutions. I re-enforced the need for some acoustic treatment in a lounge. Then, we looked at speaker technology: powered speaker settings, wave width, tweeter issues, a mix of aspects in speaker design, and looked at sweet spot, and the role of dialog clarity, usually for movie watching. In terms of stereo hi-fi music, we learned about sound stage and imaging, then the importance of subwoofers. Beyond that we explored surround sound playback technologies, and how Atmos can be added on top of that. The key Aspect of Atmos is adding height speakers, to create a fully immersive experience. I point out that this immersive experience has long been considered, not just for movie experience, but also Hi-Fi, in upward-firing ortho-acoustic speaker design, which has been a mainstream approach in Sweden. On that, I briefly explained my own ortho-acoustic speaker build.

CHAPTER 12
EQUALIZATION AND HOME LISTENING

'I'm always tweaking, always trying to make it better, constantly moving the levers and dials", by Steve Ells.

EQUALIZATION, often referred to as EQ, compensates for speaker and room (mode) inadequacies, thereby reinstating the original audio, rather than a skewed version of it. But, it involves compromise because of the artifacts the EQ digital processing causes, and so can create bigger issues if not applied in the right way. And, it is localized, meaning it only fixes the sound in one listening spot or at a two-people-snuggled-close listening spot in a room, thus skewing sound wrongly elsewhere in the room. If that room is a pair of headphones, then great; if it is a room, it is a final tweak, to be careful with.

In short, EQ is beneficial when applied to:
- headphones.
- subwoofers via AVR automated room calibration software.
NOTE: It is better to focus on room acoustics and multi subwoofers.

It is important to understand that room acoustic treatment is a first stage of EQ, one that improves the whole room, and so is far superior than electronically applied EQ because EQ only resolves skewed frequencies in one place in a room, skewing it even more elsewhere. Room acoustic treatment improves the whole room and reduces the need for electronically applied EQ, which has a smothering effect on sound, anyway. But, in this chapter explore it because it does offer some part solutions.

INTRODUCTION TO EQ AND ROOM CALIBRATION

EQUALIZATION (EQ) AND 'ROOM CALIBRATION' uses EQ to compensate for lower inaccuracies in sound reproduction through speakers in rooms, and the most common application of this if for AVRs which prioritize movies, not music playback. For music, analog purity matters more, in that case EQ is mostly not pursued. Soundbars, however, are not so good to EQ, because of the limited size of the speakers.

LET US FOR A MOMENT ESTABLISH A BASIC UNDERSTANDING of what we are working with, when we are talking about room calibration. Below is a **frequency response graph,** showing the frequencies of a recorded sine wave sweep in my lounge listening room. To measure in a preferred listening position, a sine wave is played from the speakers (aka sine sweep, see www.audiocheck.net/testtones_sinesweep20-20k.php). It moves from 20Hz to 20,000Hz (20kHz) at the same volume. The frequencies are represented along the bottom of this frequency response graph, from 20Hz to 20kHz, and the volume level on the Y axis:

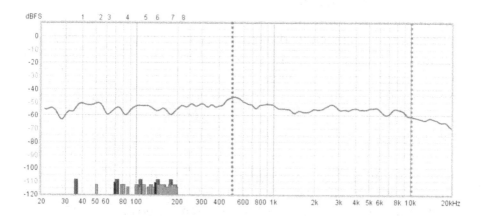

The line shows how evenly, or not, the device being measured is sounding in a particular place in a room. If the sine sweep was represented 100% accurately without any room mode interference or inaccuracies in the speakers and microphone, the blue line would be dead flat. To the degree the blue line wavers with peaks and valleys at different frequencies, a combination of the speakers and room and microphone are inaccurate in terms of volume at different frequencies. In the above graph the degree of variation is very normal and actually a very good result – because this room already have a reasonable amount of acoustic room treatment.

In the graph, there are some bars along the bottom which represent expected room modes under 250Hz, governing the dimensions and size of the room. These can be instructive in seeing whether acoustic treatment has tamed those room modes. So, if there are large peaks at those lower room modes, you know you need more bass trapping, or more people in the room, whose bodies have the same bass trapping effect. If there was no acoustic treatment it is likely the variation in the peaks and nulls would be 20dB or more; as it is, the variations are more within 10dB, which is as good as you could expect in a lounge.

The first dotted line indicates the 500Hz mark which is as far as EQ should be applied. It is being too fussy and inaccurate to EQ above that. Above 10kHz should definitely not be EQed because the measurements start to represent the inadequacies of the measurement microphone.

In order to balance out the frequency volumes, you use EQ software to reverse-mirror the peaks and troughs, determined by EQ settings of frequency, gain + or -, and the broadness of the curve (known as Q). Basically, if there is a +10dB volume peak at a certain frequency, at a certain place in your room, as caused by your room, if you EQ apply a -10dB it evens it out at that place in the room - so the listener has to sit in that location to benefit. In fact if you apply EQ as just explained and then sit a foot to one side, the room might measure a dip there instead of a peak. So room calibration is only useful for a listening position or two people snuggled together on a couch.

It does, to my mind, make the whole idea of EQ pretty questionable for a listening environment. To get the benefit, you have to sit in only one position, if you or anyone sits anywhere else it is probably worse than not having it, with the EQ degrading the quality of the audio for the whole room. So, you have the issue of needing to turn EQ on or off depending if it is just one person listening or more around the room. Then the processing of the EQ adds noise, like a blanket over the sound, another tradeoff. It just isn't worth the trouble, I feel.

So, to me, there are only three really useful applications of EQ. One is for audio production in order to make proper decisions when mixing and mastering, as discussed in the next chapter. The second if for listening with headphones. The third is to EQ subwoofers. First let us look at EQ for headphones.

▨ EQ FOR LISTENING WITH HEADPHONES

EQ FOR HEADPHONES IS A GREAT IDEA, because headphone cups (also open cups) create a room, always in the same location, over your ears. This means applying EQ generally works very well for adding clarity to headphones. You can try it out and decide for yourself about the noise smothering tradeoff. EQing headphones is easy because you do not need to measure them - there are calibration files for all common headphones available online.

WE DO NOT NEED TO MEASURE HEADPHONES, but so you know, headphones can be measured with dummy ears, see www.minidsp.com/products/acoustic-measurement/ears-headphone-jig. This has already been done for most models of headphones, and the EQ calibration details are available, for free online (as explained below).

EQ for headphones connected to our phone while on the move:

FOR AN ANDROID SMARTPHONE, you could use the Wavelet App, see www.xda-developers.com/make-your-headphones-sound-better-automatic-eq-wavelet/. However, in my opinion it is messy to deal with EQ within software apps.

A BETTER OPTION, for iPhone or Android, is a hardware headphone DAC/AMP, whereby you plug your headphone jack into it and then that into your playing device. At the same time as improving audio quality, you can also add the EQ for your particular headset. Again, you can easily get the EQ settings for your headphones from github, as linked to above.

For mobile use, I recommend the Qudelex T71 DAC AMP [32-bit/864kHz, which has 20 band EQ], see https://www.qudelix.com/products/copy-of-qudelix-t71-usb-dac, and which works for iPhone or Android:

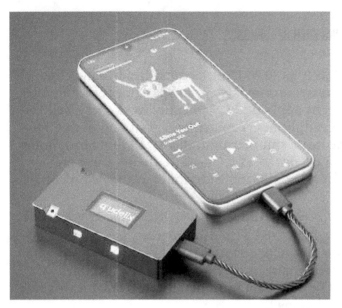

The iOS/Android mobile APP that goes with it, allowes you to enter the detailed EQ as per your headphones. It is an important feature for a mobile DAC if you are happy with the tradeoff. Here is where you enter the EQ data you get from github, see jaakkopasanen's provided data at https://github.com/jaakkopasanen/AutoEq/tree/master/results for your model of headphones.

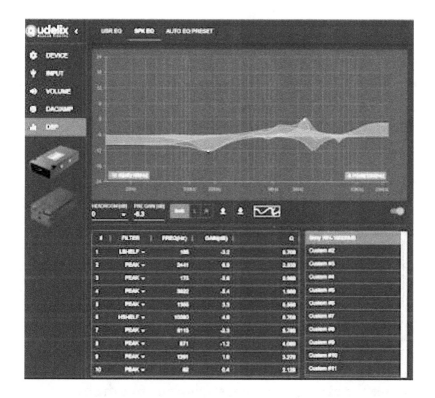

EQ for headphones at home.

WHERE YOU HAVE A GOOD HEADPHONE DAC/AMP AT HOME, you will have other options.

Software Media Players like JRiver, Roon and Audirvana accept VST plugins. I run music through JRiver on my laptop, and have used **SoundID Reference** from Sonarworks (MacOS and Windows), see www.sonarworks.com/soundid-reference, using it as VST plugin in JRiver > SSP Studio:

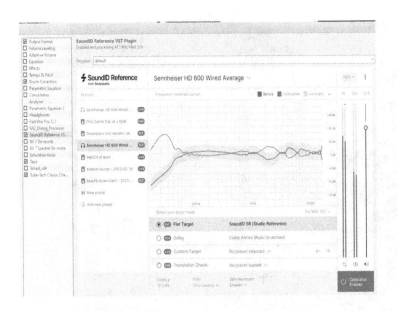

You can see at the top, the heading is 'Sennheiser HD600 - these are my headphones, and so SoundID Reference provides an EQ profile for them, as you select in the software (no need to go to github).

You can see the extreme erroneous adjustment of the bass. That is an awful lot of increase, not accounting for the real inability of the headphones to play bass all that well below 40Hz. A huge EQ jump as suggested can then cause artifact noise issues.

In fact, I had added a bass modification to my headphones, which in itself brings the bass up, meaning this curve will not be accurate anyway. Therefore, I customize the profile for my HD600 headphones by dropping the EQ bass so it makes no difference in the EQ settings, and I roll off the top end slightly, which is a normal listening preference, by the way. You can read about the modification I made to my headphones at the end of the chapter on 'Headphones, DACs & AMPs':

However, I do not do this now because the processing does take some of the life out of the sound.

ANY VST PLUGIN can be used in Roon, Audirvana, or JRiver on a PC or Mac. In fact I found the best EQ I could get was by using a plugin called Fabfilter Pro Q3. It is a very clean (minimal artifact noise) well known plugin used in the audio editing and mastering industry. See Fabfilter EQ-3 VST plugin, www.fabfilter.com/products/pro-q-3-equalizer-plug-in. I simply entered the EQ profile for my HD600 headphones from github, see github.com/jaakkopasanen/AutoEq/tree/master/results. Likewise, with SoundID Reference, I voided the suggested bass adjustment to compensate for the Headphone modification I had done, and rolled off the top end a little, as a listening preference:

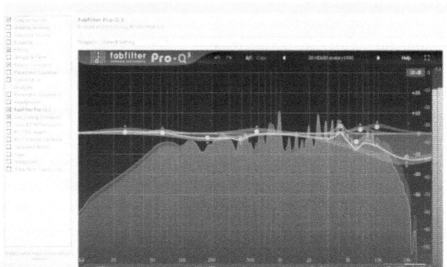

While the sound is cleaner, for listening, one still has to ask if the life is being taken from the music – try a test on and off, listening for the visceral liveliness of music, not cleanness, and see what you think. Part of Fabfilters secret is a Natural Phase option to select when processing:

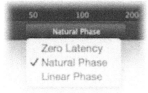

 EQ FOR HI-FI – THE LIMITATIONS

AUDIOPHILES are those people who are more consumed than others by listening to pure audio, being motivated by sound quality. I guess that is me, on the budget end. Listening to music is their hobby, and that naturally leads them to think carefully about the gear they use, and perhaps spend more on it.

Many Audiophiles shy away from EQ for lounge listening in

favor of quality gear, speakers and acoustic treatment. I do agree with this. Here is a critical and scientific reflection on the limitations of EQ and DSP, by Rod Elliott of Elliottt Sound Productions, see sound-au.com/articles/dsp.htm. The issue goes like this: quality digital interfaces and DAC/AMP use DSP chips. The initial mandatory DSP processing function is to recognize noise created by the DAC and filter it out – this is good. Then, the DSP chip, usually, enables other feature controls, such as EQ, which, if used, applies further processing. The question is whether this further processing makes things better or worse, and the question is whether any EQ settings you apply are correct in the first place. Apart from this problem there is also a limitation in accurately measuring a room for EQ. For headphones, yes; for a room, I think not, but rather work on room acoustic treatment.

On the measurement accuracy issue, we already know that in a room we are dealing with flat surface reflections. Sound reflection means we are hearing the same sound arriving at different times at our listening position. Any room measurement microphone will be summing what it hears, which includes picking up direct reflections as well as direct sound. This inaccurate information, usually summed by multiple placements, is used to decide on an EQ curve. This is why acoustic treatment is always a better solution, because it goes towards removing the direct reflected sound arriving at our listening or recording position in the first place. Our ears (or rather our brains) also have a role in dismissing out-of-time reflections, so in a way our brains are already performing a kind of EQ.

In a lounge listening space it is unlikely we can apply comprehensive acoustic treatment, which would make EQ room measurement accurate. This is especially true in modern houses, where large windows and ranch sliders, and open-plan (very reflective) kitchens are common. While an audiophile will probably go to the trouble of applying as much treatment as they can, and invest in the direction of quality speakers and amplification, and a quality DAC (for digital to analog conversion) reflections are arguably still going to make microphone measurements inaccurate. Therefore, and for the reason of EQ processing noise, EQ is arguably best left alone if purity of audio is your goal, especially if two or more people are listening in different chairs, which totally nullifies EQ since EQ is only for one listening position, and skews sound elsewhere in the room.

I therefore come down on the side of not using EQ for Hi-Fi listening. I think we are better off focused on adding acoustic treatment, upgrading our gear, and always inviting others in to enjoy the audio. But, subwoofers are perhaps an exception, if you sit in one place or snuggle on a couch with another person, the trade off between the processing noise of EQing the subwoofer only and improvement is probably worth it on the proviso you do it through an AVR where the subwoofer is assigned the low-frequency effects (LFE) channel.

I do admit, for Hi-Fi, it is hard not to try it out, because the question does arise as to whether you loose more from some signal noise compared to what you gain by EQing. It is a tradeoff and the gear you have may impact that. In the first case, I suggest trying EQ out if you have an AVR, because it is automated and when watching movies on your own or snuggled up, the trade off is probably in favor of doing it. Though, I still personally prefer putting the effort into acoustic treatment. I want to minimize suffocating audio with any unnecessary processing because it relates the visceral connection to the sound, and thus the experience of listening whether it is music or a movie.

EQ TV AND A SOUNDBAR

I WILL NOT RECOMMEND THIS, and I add to my reasoning. If you get a Soundbar it may have an auto-EQ/room calibration option. I concede, maybe some of the better ones can benefit from EQ, on the proviso that you have some room acoustics in place and you listen from one position. Here is a Samsung video for optimizing sound with SpaceFit Sound, which is an automatic EQ calibration setup, for a Samsung Soundbar with a Neo QLED TV, see youtu.be/EwTESGG85dQ. I would expect these features to become more standard with newer high quality Soundbars. And, the Subwoofer in this Q Series system, will EQ automatically, see www.samsung.com/nz/audio-devices/soundbar/q990b-black-hw-q990b-xy/#SpaceFit%20Soundplus:

SpaceFit Sound

SpaceFit Sound offers clear and immersive sound by optimising your audio to match your space. The soundbar can even automatically analyse the room on its own, calibrating the sound field to perfectly fit your room.

Auto EQ

The included subwoofer automatically calibrates to provide any space with powerful bass.

IS IT WORTH IT? For most Soundbars, I think not. If you tend to have more than a few people watching TV, forget room correction EQ because that only works in one place (or for two people snuggled up on a couch). Secondly, EQing like this across the frequency spectrum is only effective if your gear is really good, and your room is acoustically treated. I have previously explained that one of the intrinsic limitation of Soundbars is the smallness of the speaker cabinet. Taking these factors into account, I would not suggest applying EQ for Soundbars.

 EQ AVR

 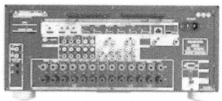

WE HAVE SLIGHTLY BETTER CONDITIONS WITH AN AVR. It is a more expensive system, so it is more likely you will have quality speakers. As long as you have a reasonable amount of acoustic treatment, and you are a solo listener, or two, snuggled on a couch, then, it should work out well to use the auto EQ, setting it for up to 500 Hz. Most AVRs come with auto-EQ software. Yet, I still claim you will loose some visceral connection – try it out though.

 SUBWOOFERS IMPROVE TONAL CLARITY

At www.spinorama.org/scores.html?quality=High you can compare the scores for different speakers, and also see that there is almost a 10% tonal improvement with a subwoofer added:

Brand Model	USD	Reviews	-3dB	Flat.	Tone	w/eq	w/sub	w/both
KEF LS60 Wireless	3500$	KEF	23.4Hz	±4.3dB	8.0	8.1	8.6	8.7
KEF Blade 1 Meta	17500$	KEF	34.3Hz	±1.1dB	7.7	7.8	8.6	8.9
KEF Reference 5	8400$	KEF	31.3Hz	±3.9dB	7.6	7.7	8.5	8.6

This is a good argument to add multiple subwoofers to your system in the first place. Not only are they more responsive to EQ, but multi-subs complement each other by meshing and equalizing sound out in the room.

 EQ ONE SUBWOOFER ON A STEREO SYSTEM

AUDIOPHILES DO NOT LIKE EQ, as I have said, because it is a process added to the digital file that will add some noise, even if slight. Sometimes it requires analog to digital conversion so as to apply the EQ, then back into analog, and that is after the main DACs digital to analog conversion.

However, if you have added a separate subwoofer to your system to improve low bass performance, I think applying EQ just to the subwoofer is a good idea. You will get a noticeable improvement in clear bass notes, and less destructive influence on the harmonics to the tone higher up. If the subwoofer is separate you can add a miniDSP device on wires to the subwoofer, thus having no impact on the main audio signal.

If you have a stereo system (not an AVR), and so do not have auto-EQ, then you can add a miniDSP device pre-connected to the subwoofer, thus having no impact on the main audio signal.

I suggest connecting a DDRC-24 between your power amp and subwoofer, see www.minidsp.com/products/dirac-series/ddrc-24. The EQ auto-calibration will be easy to set up because it has Dirac Live. Just set it to measure the signal just past your crossover. If, for example, you crossover at 80 Hz, set it to measure 20 Hz to 100 Hz.

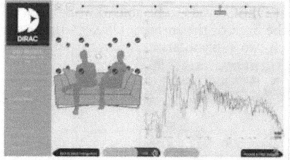

MiniDSP also offer a miniDSP IR remote, see www.minidsp.com/products/accessories/ir-remote, to select EQ profiles depending on where you are sitting, or you can turn EQ off if there are people listening around the room:

EQ MULTIPLE SUBWOOFERS

Also refer to 'Subwoofers' and 'Subwoofer setup' sections in chapter 11 'Lounge Listening'.

PREFERABLY HAVE TWO OR MORE SUBWOOFERS, because they act like room acoustics, specifically, to produce all the frequencies between them – no nulls in total. As long as you EQ them, you will get balanced bass and then you know no destructive harmonics are influencing the tone higher up.

FOR WATCHING MOVIES OR TV, WHAT YOU REALLY WANT is an AVR compatible with Dirac Live Bass Control (DLBC). You can check for DLBC compatibility at ww.dirac.com/online-store/, as with the Intregra DRX-9.4, Denon AVR-A1H, and others. This will handle multi-sub EQ setup in an automated step through process. Also, check how many Pre outs for subwoofers there are. These are independent outputs enabling separate controls for each subwoofer, which is what DLBC needs to be able to work with. Here you can see from the back of the Intregra DRX-9.4 that there are two, so you can attach two subwoofers:

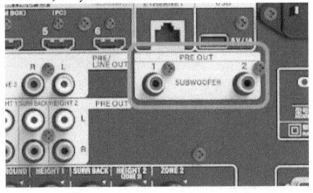

Here with the Denon AVR-A1H, there are four Pre outs for subwoofers, using either RCA and XLR   connections. Therefore, this would allow for four subwoofers, for DLBC to set up:

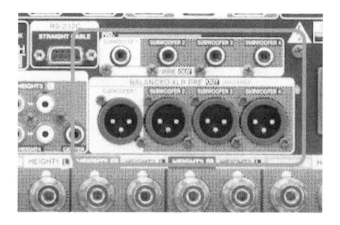

YES, YOU CAN EQ MANUALLY, but you still need an AVR with multiple sub outputs, but it is quite technical with a number of conditions. DLBC is the solution we have been waiting for. Perhaps other EQ providers will come out with something similar, but as of June 2023, DLBC is on its own.

SUMMARY

I BEGAN by stating the limitations of Equalization, being reminded that room acoustic applications reduce the need for it, even entirely in the case of a recording studio. Since EQing headphones is effective, I explained different ways that is done. In a lounge listening environment we only want EQ for subwoofer, I believe, but even then multiple subs are a better all room improvement. Really, if we are listening to Hi-Fi music, we probably do not want EQ, because in itself the EQ processing adds a certain amount of noise.

I have also suggested steering away from EQ for Soundbars because of limited speaker cabinet size. They simply are not up to the quality audio level for EQ to be effective, governing room reflections being part of the measurement issue. Since, as we have said, EQ is relatively effective for Subwoofers, I delved into that, including multiple subwoofers. But, I point out that multiple subwoofers benefit from a very good quality AVR with separate subwoofer connection points, and auto EQ through the new Dirac Live Bass Control (DLBC) system.

A Subwoofer, can also be added to stereo Hi-Fi systems and EQed, but in that case we need to know how to EQ just the subwoofer using a miniDSP devise. In the next chapter, I swap over to Audio Production and EQ on that side of the fence.

CHAPTER 13

AUDIO PRODUCTION AND EQ

"Tone is always such an important thing, and that's achieved through a multitude of people. It comes through the writing, it comes through the way it's shot, and it comes through the production design and the sound design", by Harry Treadwell.

IN THIS CHAPTER, we will discussed EQ in the post recording stage, how EQ is applied to resolve room mode tonal interference from mixing decisions, and also by way of tonal effects. For the home-based producer, I will introduce auto-EQ plugins, offering what I believe is an improvement over methods like EQ sweeping (also discussed in this chapter). I will also cover standard recording presets and the two main editing techniques of comping and punch-and-roll. Again, there is a system of parts that work together, or I could say, a series of steps, each needing to be undertaken in an effective way.

AUDIO PRODUCTION AND A STARTING POINT FOR EQ

Maybe you should employ an audio engineer

IN THE POST-RECORDING STAGE, each track will need slightly different applications for acoustic and tonal effects (which are variation of EQ). In my case I spent a lot of time trying to apply EQ to audio dialog. It was an investment of time, not having the fully trained ears of an audio engineer. If you have the financial resources, a worthwhile approach is to prepare your tracks through the editing process, then send them to an audio engineer to do the mixing and mastering work. That will be a huge time saver and allow you the time to move onto your next project a lot sooner. Unfortunately, I did not have the budget for that approach. One way to locate mixing and mastering services is through Fiverr, see www.fiverr.com/categories/music-audio/mixing-mastering:

Try auto-EQ plugins

I FOUND THE EASIEST WAY for me, including see below on headphone calibration, to sort out EQ, for post recording listening decision clarity, was to rely on the auto-EQ plugin, smart:EQ 3, from Sonible, see www.sonible.com/smarteq3/:

It is a very good way, to create an EQ starting point, then you see if you can tweak it slightly using the A/B comparison controls - you always have the smart:EQ 3 auto-generated base line to fall back on. You might use the vocal profile, or the one for Bass, Keys, acoustic guitar, electric guitar, drums, kick, snare, etc. This means you can apply an instance of the plugin for each different track, run its automatic EQ depending on the instrument, then, as you wish, see if you can tweak those settings. After doing that, I did reproduce the EQ curve in Fabfilter Pro-Q 3 because it ended up sounding cleaner, and after doing so, I deactivated smart:EQ 3. To tweak settings with some understanding , a good way is to search Youtube for examples: EQ for dialogue, EQ for vocals, EQ for electric guitar, etc. Here are some other educational videos I liked: EQ aspects of a voice, by Colt Capperrune, see youtu.be/Wq1di2luMcs, and Colt Capperrune's use of Fabfilter Pro-Q 3 for the audio part of podcast video recording, see youtu.be/hvDKeG63Qik.

EQ - live venues

EQ (meaning room calibration EQ) ON THE RECORDING INPUT PATH can be useful for a live stage scenarios. For that scenario, there are tools like the dbx's DriveRack, see dbxpro.com/en/products/driverack-pa2. It has an auto-calibration step-through system (using a reference microphone). In terms of acoustic treatment in a venue, human bodies work well. If there are 25 people in a venue, that is equivalent to a 100 square foot of wall covered with 4 inches of fiberglass. So, you should take the measurements when the expected audience is in the room.

HERE IS AN ARTICLE FOR MICROPHONE PLACEMENT, AND SELECTION, FOR A LIVE VENUE environment, by Sweetwaters, see www.sweetwater.com/insync/best-live-sound-mic-placement-techniques/.

 ## ROOM EQ PLUGINS IN A DAW

ONCE YOU HAVE RECORDED AUDIO, you will be applying some effects such as EQ, compression, and Reverb. But, in order to make informed decisions, you need to remove your own localized environment, meaning any inaccuracies of your headphones, monitors/speakers and how your room modes may twist the sound. When you do this the audio you hear is the same as the actual audio media. Your new EQ compensation setting will be in an EQ plugin in the Post section of your DAW (see red highlight, below). When you finally output a file, you turn these off, because they are only needed for your listening as you edit.

To get the necessary EQ data, you can use the free online EQ data, as linked to in the previous chapter, at github, for your headphones. As discussed, you can use any EQ plugin, and I prefer Fabfilter EQ 3 because it is so clean. For your monitors, you will need to measure them. One common tool is **SoundID Reference** from Sonarworks (MacOS and Windows), see www.sonarworks.com/soundid-reference, using it as VST plugin in your DAW. There is a version for headphones only and another for headphones and speakers, which is the one you would probably want. You could equally use Dirac Live for Windows, see www.dirac.com/live/studio-audio-for-creatives/. Let's first check the EQ for hi-fi speakers with Sound ID reference and their measurement microphone, see store.sonarworks.com/products/soundid-reference-measurement-microphone. SoundID will step you through a measuring procedure to produce EQ profiles for your monitors. The results of the measurements will be something like this:

EQ for my Mackie 524 *monitor speakers:*
The **purple are the measured variation of sound from flat** (flat meaning an accurate representation of the original audio), for both monitors. The **green are the EQ suggested frequency changes** to balance it.

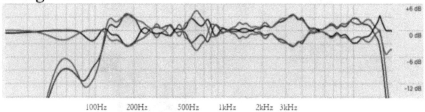

Both speakers are pretty much the same, three main issues being at: 180 Hz, 600Hz, 3.5kHz. Making these changes will improve the Mackie's audio output accuracy significantly, enough to reliably work with audio. Likewise, with the headphone VST settings, the monitor/speaker EQ compensation will also go in the Post area (see above).

SETTING UP TO RECORD A TRACK

BEFORE RECORDING THERE ARE A FEW SETUP REQUIREMENTS. You may not need to know this, but I though this was the right place to slip this in. First, here is a useful video on setting up your Digital Interface, by Audio University, see youtu.be/u_qCh3TIk8w.

PRESET RECORDING VOLUME: After deciding on the distance of your mouth or instrument to your microphone, usually a fingers-spread distance for vocals, a few inches for musical instruments. In a DAW you need to make a track live, then set the gain on your Digital Interface (DI). On your DI there might be a flashing indicator around or near the volume knob, indicating how the recording level is being picked up. Make the highest sound you are wanting to record (governing your microphone distance, and movement of a singer*), adjusting the knob up until you see a red flashing light, or your indicator jumps up to red. Now, adjust it down a bit, so with the loudest noise you see yellow, but not red, This setting provides the lowest artifact noise level in relation to the audio you are recording, otherwise known as the 'Signal to Noise Ratio (SNR)'. If you turn the gain knob down more than this, the SNR will be worse; you will have more artifact noise in relation to your desired recording, not good.

* Sometimes a singer or narrator will get really close to a microphone for the proximity effect it provides. The artist should be able to control their voice by lowering it when they do this. Whatever their habit, you do not want volume recorded into the red where it clips off the top of the recorded sound, not good.

This recording Digital Interface (DI) gain indicator will be reflected in your DAW's Input signal meter. As a rule of thumb, the method of setting above should give you a maximum average recording level of around -18 to -12 dBFS.

By the way the Input meter in your DAW should not be confused with Output meters. The following video by Home Recording Made Easy, illustrates this point, and showing how you might have multiple input meters for a digital interface with multiple microphones attached, see youtu.be/Qz0afhER6fo.

If you observe the Input meter you should be seeing, average peaking at about -6 dB, some peaks up to -3dB or just above, and no clipping (hitting the red line of 0 dB). The track you record should look something like this:

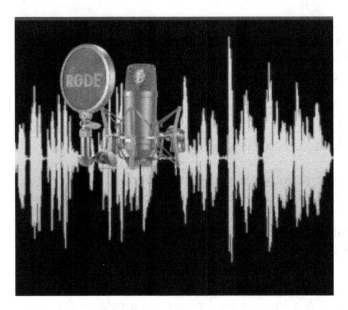

If you measure the average RMS of a track, it will probably be between RMS -18 to 23 dBFS, similarly if you are measuring in LUFS, though that will show maybe 2 dB lower in volume. RMS is the average voltage of the signal, which represents volume; LUFS, a newer measurement approach, is calculated based on how we perceive sound with our ears. One good way to measure your Input volume in detail is in Izotope RX, see www.izotope.com/en/shop/rx.html, which I refer to a number of times as an indispensable tool for Audio recording. In RX when you are looking at any track, you can select Waveform Statistics:

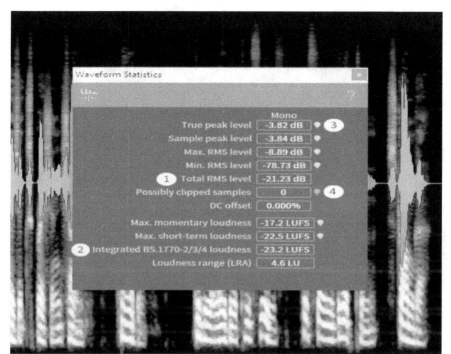

(1) shows the RMS level, which we have said should be between RMS -18 to 23 dBFS. (2) shows the LUFS level, which we have said may show a couple of dB lower. (3) True peak shows whether you would clip the audio if you saved the audio to a final file format, and (4) tells you whether you have any clipping in the file – this should be zero.

A couple of other useful tools that you can use in your DAW are the Free Voxengo Spectrum Analyzer Plugin, see www.voxengo.com/product/span/ and the cheap WLM Meter from Waves, see www.waves.com/plugins/wlm-loudness-meter. These are VST plugins, which you turn on for analysis purposes, such as Gain Staging where you check volume is the same in and out of your effects plugins. Drop them on your track after a recording to analyze the RMS and LUFS. Here (1) you can see the RMS readout on the Voxengo Spectrum Analyser Plugin:

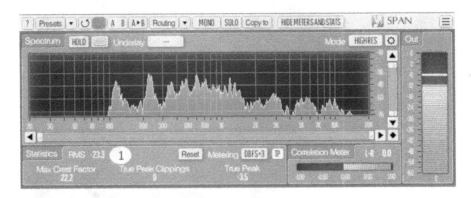

Here (2) you can see the LUFS readout on the WLM Meter:

Bear in mind, we are not talking about the RMS or LUFS measurements of a mastered file ready to supply as a finished product. We are looking at optimal recording settings on your Digital Interface, in order to minimize the artifact noise levels that you will inevitably have, but want to minimize.

 RECORDING & EDITING

COMPING, AND EDITING is when you you take the best parts of multiple takes and edit them together. Since this can get very difficult, for example, The distance between the instrument or voice to the microphone might be slightly different with two tracks, and they won't sound exactly right. It is far better to keep recording multiple tracks (retakes) until you get one that seems perfect. Then as you go through it you might find just a few issues, and for those, you have a choice to go back and rerecord or take sections of other recordings and splice them in. It will always be better to have multiple recordings at hand that have been made with the same microphone, in the same conditions.

PUNCH AND ROLL, AND EDITING is a recording technique more common for long form dialog recording. It is a method where, when you make an error you pause your recording, and punch in before the phrase where the issue starts. You will hear the phrase before and the track image shows where you need to start talking. You keep doing this, fixing errors as you make them, thus your editing has been done as you record. Here is a video of this process, by Don Baarnes, of Red Baarnes Audio, see youtu.be/pzhWImZDlV0.

AUTOMATED DIALOG REPLACEMENT (ADR) is commonly used to fix audio tracks for video. This is more difficult, taking background sound into a track, then recording in a studio with the right amount of resonance (distance from microphone) with the same microphone used for the Video. This recording should be made to fit the timeline for what you want to replace. Some EQ can be applied to get the sound even closer to what else is on the track. Here is a video showing the process, by Shutterstock Tutorials, see youtu.be/7W46DjoFlGE. This will be even more difficult if you are trying to lipsync. In that case you might use a plugin like Synchro Arts VocAlign Ultra, as in this tutorial, by Production Expert, youtu.be/Sbszd20p8-M. In this case you still will want to have recorded as close as you can to the original dialog, in gear, in the same place (or as close to is as possible), in timing, in tone, in expressiveness.

 ## EQ SWEEPING FOR MIXING AND MASTERING

EQ SWEEPING, is commonly used for mixing and mastering, in post recorded audio, to remove resonant frequencies, which are caused by room modes. You can hear the modal/peak issues using an EQ plugin if you create a very narrow raised EQ band and sweep it back and forth over the frequency spectrum of your audio, until you hear a shrill- whistle sound. Then you can EQ sculpt that area a bit, counteracting the issues.

I do not really like this system, because it relies on your ears rather than accurate measurement, but it is commonly used by audio engineers, who, do have trained ears, while not having access to the original recording environment.

The best way to understand EQ sweeping is to see a few YouTube videos of it. Before doing that, just a word about energy. A volume peak is more energy, and a null is less energy. In general, if you reduce energy, by lowering a peak, you should raise a null somewhere in tonal symmetry, so that the energy in the music is maintained. This question of balancing the energy is sometimes mentioned when audio engineers discuss EQ sweeping: EQ sweeping by PoundSound, see https://youtu.be/RFuHHOVYUO0.
EQ sweeping by Collaborate Worship, see https://youtu.be/bfggxWDc4Lg.
EQ sweeping by Andrew Zeleno, see https://youtu.be/CCNORzxfgb0 shows how easy it is to go overboard. Andrew explains how an audio engineer should hearing problems before attempting EQ sweeping.

I PREFER THE ALTERNATIVE OF TRYING AUTO-EQ PLUGINS, as discussed at the top of this chapter, because I am not an audio engineer by trade, and do not have developed listening to the extent they do.

Bear in mind, if an auto EQ process comes up with significant

EQ curves, lets say more than 6 dB of an adjustment, then this shows you you should improve on your room acoustic treatment and/or the recording position. Any EQ adjustment over 6dB starts to noticeably add its own artifact noise to the audio.

EQ TONE SHAPING FOR MIXING AND MASTERING

ONCE YOU HAVE RECORDED A TRACK, OR TRACKS, in order to edit, mix and master there are a number of plugins you will use, whether these are proprietary ones in your DAW and/or VST plugins you purchase.

You can get lost in the maze of plugins available until you settle on ones you are comfortable with for the sound you want. This is where sending edited data to a professional for mixing and mastering can be a really good option, saving you a lot of time and effort. Editing, though, is the first job that you would in any case likely be doing yourself.

I think a good way to keep your plugin management under control is to follow the advice of a professional audio editor's favorite plugin list, and learn those well before branching out. Here is what I am talking about, for music, favorite plugins from Produce Like A Pro, see youtu.be/EDezID2RH6g.

If that is a little daunting for you and you want something simpler, one tool that can achieve an **overall mixing process for music** is Neutron 3, see www.izotope.com/en/products/neutron.html. And, one tool that can achieve an **overall mastering processing for music** is Ozone from Izotope, see www.izotope.com/en/products/ozone.html.

MIXING IS QUITE AN ART, which I am not claiming to be a master of. There are many YouTube videos which do provide excellent tuition on it. As a starter you could watch this video on The 5 Rules of Mixing, from Produce Like A Pro, see youtu.be/NE1Ov-ObLv0.

BEFORE CONSIDERATIONS TONE SHAPING, we can consider the right kind of microphone we should be using depending on the instrument we are recording. In the chapter 'Microphones, Digital interfaces, and Plugins', I have introduced the main kinds of microphones, for a basic understanding. What I did not do in that chapter is go into which specific brands and models suit certain instruments. In the circuitry and engineering design, different microphones can be better for certain instruments and types of music. I think the best way is to find recordings of material similar to what you are working on, and try and find out what microphones were used. That will be a reasonable guide, though for vocals there may be more variation regarding matches to particular voices. This process, will reduce the amount of tone shaping you may want to do when mixing and mastering, but it does not take a way the need for it in post production. Here is one article discussing different matches of microphones to music, by Live Toolkit, see www.livetoolkit.com.au/guide/microphones. And another by Sweetwaters. Although the article is about microphone placement, is very much about microphone choice for different instruments, see www.sweetwater.com/insync/best-live-sound-mic-placement-techniques/.

A prerequisite of mixing is to understand **tone shaping for mixing and mastering.** Having considered the right microphone for the job, further tone shaping is often applied after recording. The process of **tone shaping a** track is, firstly, to highlight the strengths of the instrument tonally and vice versa for other tracks. Secondly, you may carve out EQ dips in one track so another track will shine through (suitable for stereo recordings, not for surround sound). Here is an explanation by recordingrevolution, see youtu.be/uoJqNzxYQD4. Here is a pretty good Tonal Shaping Cheat Sheet for different instruments, including the human voice.

Instrument	Frequency to Cut	Frequency to Boost
Human Voice	7 kHz: Sibilance	8 kHz: Big Sound
	2 kHz: Shrill	3 kHz and above: Clarity
	1 kHz: Nasal	200 - 400 Hz: Body
	80 Hz and below: Plosives	----
Piano	1 - 2 kHz: Tinny	5 kHz: More Presence
	300 Hz: Boomy	100 Hz: Bottom End
Electric Guitar	1 - 2 kHz: Shrill	3 kHz: Clarity
	80 Hz and below: Muddy	125 Hz: Bottom End
Acoustic Guitar	2 - 3 kHz: Tinny	5 kHz and above: Sparkle
	200 Hz: Boomy	125 Hz: Full
Electric Bass	1 kHz: Thin	600 Hz: Growl
	125 Hz: Boomy	80 Hz and below: Bottom End
String Bass	600 Hz: Hollow	2 - 5 kHz: Sharp Attack
	200 Hz: Boomy	125 Hz and below: Bottom End
Snare Drum	1 kHz: Annoying	2 kHz: Crisp
	----	150 - 200 Hz: Full
	----	80 Hz: Deep
Kick Drum	400 Hz: Muddy	2 - 5 kHz: Sharp Attack
	800 Hz and below: Boomy	60 - 125 Hz: Bottom End
Toms	300 Hz: Boomy	2 - 5 kHz: Sharp Attack
	----	80 - 200 Hz: Bottom End
Cymbals	1 kHz: Annoying	7 - 8 kHz: Sizzle
	----	8 - 12 kHz: Brilliance
	----	15 kHz: Air
Horns	1 kHz: Honky	8 - 12 kHz: Big Sound
	120 Hz and below: Muddy	2 kHz: Clarity
String Section	3 kHz: Shrill	2 kHz: Clarity
	120 Hz and below: Muddy	400 - 600 Hz: Lush and Full

This cheat sheet is provided by Presonus, along with other guidelines, see www.presonus.com/learn/technical-articles/How-Do-I-Use-An-Eq. And, then it can get more detailed, for example, this information on Vocal EQ Cheat sheet, 343labs.com/vocal-eq-cheat-sheet. Fabfilter Pro-EQ 3 or other EQ plugins can be used to achieve this tonal balancing.

 ## MY PLUGINS FOR VOICEOVER

THESE ARE THE PLUGINS I HAVE USED for Audio editing and mastering my dialog recordings.

RECORDING:
Studio One Limiter (catch any possible clipping at -1dB when recording, so on the input effects area in the DAW. It will round any clipping off as it comes in. Though make sure the recording level is set to avoid it) [blog.presonus.com/2019/03/29/friday-tips-limiter-demystified/].

CLEANING:
RX De-Plosive (a dePlossive to eliminate any potential "p" and "b" plosive sounds that could have gotten through in spite of the microphone pop filter) [www.izotope.com/en/products/rx/features/de-plosive.html]
Fabfilter Pro-DS (a deEsser to eliminate any "Sss" sibilant sounds, that might come through) [www.fabfilter.com/products/pro-ds-de-esser-plug-in]
RX Voice DeNoise (One of the RX plugin modules to reduce hiss, for clarity) [www.izotope.com/en/products/rx/features/voice-de-noise.html]

EQ:
Smart:EQ 3 (auto-eq settings from the profile, which I may tweak slightly, to decide my EQ settings. After that I duplicating to Fabfilter Pro-Q 3 and muted Smart:EQ 3) [www.sonible.com/smarteq3/]
Fabfilter Pro-Q 3 (I duplicate the EQ settings of Smart:EQ 3, then I Disabled Smart:EQ 3. Also for tone shaping) [www.fabfilter.com/products/pro-q-3-equalizer-plug-in]. If you are

more confident, you might do an EQ sweep, but that is best if you can hear the issue in the first place, as discussed above.

COMPRESSION *:
WaveriderTG (like a compressor, and works well for dialog, that does not change the tone, like a compressor does, but brings the volume up and down for clarity) (quietart.co.nz/waveridertg/)
TDR Kotelnikov GE (upward compression, for easy listening in a car, and some tonal manipulation, especially the Yin Yang settings. I use it to avoid downward compression, which can affect tone. I like the Yin Yang settings) [www.tokyodawn.net/tdr-kotelnikov-ge/]

CLARITY and TONE SHAPING:
SA2Dialog Processor (EQ, for dialog tonal clarity. A good idea if needing to bring speech or vocals forward in a mix) [www.mcdsp.com/plugin-index/sa-2/] An alternative with more of a musical focus is **Gullfoss** (EQ based on audio perception model, for clarity, detail, and spatiality) [www.soundtheory.com/home]
Tape RC-1 (Tape coloration, for pleasing character) [www.softube.com/tape]
 NOTE: The position (in relation to the other plugins) of Gullfoss could be experimented with.

MASTERING

WLM Loudness Meter (Gain staging. A loudness meter, for keeping track of levels in and out of each plugin [www.waves.com/plugins/wlm-loudness-meter], called 'gain staging', so the same volume level is maintained through the plugin pathway. I also check RMS and LUFS in my last plugin Fabfilter Pro-L, the levels guided by the service I will send the finished audio file to, such as these requirements for submitting audiobook material to ACX, the conduit for uploading audiobook content to Amazon, see www.acx.com/help/acx-audio-submission-requirements/201456300. All upload services will have a similar page for their requirements.

Before exporting an audio finished product, make sure any test plugins and any SoundID reference or similar headphone and speaker listening plugins, that you have as Post inserts, are Disabled.

Fabfilter Pro-L (final limiter, but also loudness adjustment for outputting master files [www.fabfilter.com/products/pro-l-2-limiter-plug-in]

Not for voiceover, but if your have multi-tracks, a well know mastering plugin is **Ozone** from Isotope [www.izotope.com/en/products/ozone.html]. If that encompasses too many tools for you to want to figure out, have a look at **AI Master** by ExonicUK [www.exonicuk.com/]

> * You will notice that the two plugins I use under the COMPRESSION heading do not apply algorithms to compress the peaks, where transients are. Transients are the initial sound of consonants and the initial striking of any musical instrument. The basic idea of compression is to make the sound more audible, especially if there is background sound for the listener, by lowering high transients. Therefore, it is useful so your product can be heard in different background noise scenarios, less needed for classical music, which tends to be listened to in a home backgound-quiet environment. Beyond that basic idea, reducing transients will also seem to put an instrument into the background; a somewhat similar effect as applying reverb. There are also a number of pedal type compressors and plugins which shape transients to shape tone. On the software plugin side, try Eventide's **Split EQ**, see www.eventideaudio.com/plug-ins/spliteq/, as explained here by Sonic Academy, youtu.be/PaC8mpiQ_p4. Basically, Split EQ separates out the transients from the more sustained tonal elements, allowing for very unique tone shaping, such as to reduce transients in a frequency range so the transients of another instrument can shine through in that frequency range, in a mix, or, for example, to exacerbate the tonal

aspect of guitar strumming. It can also be used for dePlosive and deEssing processing.

Since I am working on vocals (vocal tracks), I want to leave transients alone, because an amount of the timbre/tonal color is in the peak transients. This is why I use **WaveriderTG** and **Kotelnikov GE**. WaveriderTG rides the volume, pushing down the gain of the peaks (rather than applying a squashing algorithm on the transient peaks) and increasing the gain of the quieter sounds, with a naturally sounding algorithm. **TDR Kotelnikov GE** can apply upward compression, pushing up the quieter sound, but leaving alone the construction of the peak transients. Therefore, by using these, I am protecting the part of the original timbre in the transients.

Though, some transients are problematic, such as plosives (which is why we use pop filters), and the 'S' sound can be excessive. Therefore, I do use EQ, aimed at lowering just those problematic frequencies, which I listed under the CLEANING section; **Fabfilter Pro-DS** and **RX Voice DeNoise**.

Trying to maintain the richness of original tone, without introducing digital noise is a reason for using tube-based compression in the recording chain, such as discussed in Chapter 6 Microphones, digital interfaces, and plugins..

SUMMARY

WE DISCUSSED EQ in the post recording stage, that EQ is applied to resolve room mode tonal interference, and as well by way of applying tonal effects. Resolving room mode issues may best be left to audio engineers because of the listening skills involved, though auto-EQ plugins have been shown to be useful. These, as explained, can provide template EQ adjustments defined by instrument, which then learn their way to a usable EQ profile for that instrument track. This enables home-based audio producers to pull off reasonable application of EQ.

For Audio production, many decisions are made, and so room EQ plugins in Post need to be used so those decisions are about the actual audio, not while the room is skewing it. EQ adjustments for headphones and speakers are then put in the Post section of a DAW prior to decision making. Apart from room and headphone EQ, many tone shaping (EQ like) plugins can be used – I explained some strategies so as not to be swamped by the huge number of options.

CHAPTER 14
SUMMARY

SOUNDS are produced within a frequency range, extending their harmonics, and this can be visualized in typical audio programs and effects plugins. Seeing is believing. Not just frequencies, but we can visualize the mixing and panning processes as sounds on a stage.

For recording, while certain spaces are better than others for capturing your creativity, in any case we can apply acoustic treatment to manage the sound in the room. Starting with the question of recording I illustrated an acoustically treated small space online setup (including the audio visual elements), and a studio setup.

Room acoustic applications tame room modes, and echo, to enable enjoyable visceral audio – clearer, crispier, tonally more pleasing and relaxing. The different kinds of room acoustic treatments were outlined, along with DIY instructions.

Room isolation methods were also discussed because it can be necessary to reduce the frustration of outside-of-house noise, and/or for politeness to our neighbors.

IN THE CHAPTER on Microphones, Digital Interfaces, and Plugins, I focused on how to retain aspects of analog sound. A range of different types of microphones, and how to find a sweet spot in a room to record in was covered. Analog came up again in the audiophile preferred DSP file format, and so how to record it, and Dolby Atmos. We also looked at Digital Audio Workstation software (DAWs), which led to some discussion on how to up-mixing to Atmos.

FOR HEADPHONES, we explored various types, and their various tonal, acoustic and use benefits. For DAC/AMP technology, we looked at different chip types and circuitry because their sound signatures are different, and we noted how slightly higher end gear uses technology that is more focused on an analog sound signature.

Considering cables and wires, we began by claiming the advantage of upgrading interconnect and microphone cables – it will improve the clarity and tone. Likewise, for quality speaker wire and connectors. Detail was provided on soldering, should that be needed for DIY projects. And for noise that may slip through, we saw how to hunt it down and minimize it.

For clean audio, I provided a list on how to achieve a clean recording. But it is not all about being clean and clear. The questions of analog versus digital was discussed because while some noise is bad, some noise is wonderful. We saw how high-res file formats provide cleaner clearer audio, not because the file itself is cleaner, but because of how modern DACs can process it. Well, if there is still noise on the end file, we considered how to clean up noise issues using the industry standard RX Izotope. And, back on the analog question, we explored the DSD format for its as-close-to-analog sound as you can get for a digital file, finishing the discussion by stating that it is best to keep things as analog as we can, at least in hybrid form, because it helps the emotional connection.

ON Streaming and Play-through, we saw how newer technologies are developing along with the instigation of lossless audio file formats, and what different services deliver (at the time of writing). As an example, we saw what parts make up an Atmos file format, and how it is transported and rendered.

ON lounge listening, we came to understand the listening sweet spot in a room, and the role of resonance in relation to direct sound, and the problems and solutions of speaker placement, and design. In terms of stereo hi-fi music, we learned about sound stage and imaging, and the importance of subwoofers. And, we learned about surround sound, and Atmos for which upward-firing speakers are used. On that speaker design we saw how Sweden had taken the lead, and I explained my own DIY project whereby I built four upward-firing ortho-acoustic speakers.

Equalization in a home listening environment can help, but not as importantly as room acoustic applications. And EQ is only worthwhile in a limited number of cases: AVR setups for solo listening positioning, subwoofers, and headphones; not really hi-fi.

FOR audio post-production, EQ has a definite place. First we resolve room modes and headphone inadequacies, then we add tonal EQ and effects. Audio engineers have the ears, so if you are not one, I have suggested auto-EQ plugins as an independent way forward, along with some tone shaping guidelines.

THERE IT IS, the why, what, and how of audio; the room spaces, the acoustic materials, the devices, the software, the formats, the management, the tweaking, the uses, each with its own dos and don't. Line them all up, understand them together, and you understand audio, you understand how to make your own decisions, how to record your creations and how to listen to other's.

REFERENCES

$1000 mic Shootout - Sennheiser MKH416 vs Neumann TLM 103. (2017, January 14). YouTube. https://youtu.be/sdE_VekATvE

Acoustics Insider. (n.d.). *The best (And only) insulation material you need for DIY absorbers.* https://www.acousticsinsider.com/blog/best-insulation-material-diy-acoustic-absorbers

ADAM Audio. (2021, May 20). *5.1 vs. Dolby Atmos... What's the Difference? - Dolby Atmos Masterclass Ep. 1 | ADAM Audio.* YouTube. https://www.youtube.com/watch?v=jyOOxI8-zSQ

ADAM audio. (2021, May 19). *5.1 vs. Dolby Atmos... What's the difference?* https://youtu.be/jyOOxI8-zSQ

Analog audio vs digital audio: The real difference. (2019, November 17). Audio University. https://youtu.be/844LQegRV-A

Andertons Synths, Keys and Tech. (2019, April 5). *Which ribbon mic is best for vocals? We compare three of the best to find out!* YouTube. https://www.youtube.com/watch?v=Q6FnlI4jDAU

Apple Music and Dolby Atmos on a Mac. (2021, June 12). YouTube. https://youtu.be/GUblp6EVtJ0

Arthur. (2020, February 14). *Top 11 best tube condenser microphones on the market 2023.* My New Microphone. https://mynewmicrophone.com/top-best-tube-condenser-microphones-on-the-market/

Audio for Content Creators. (2020, September 16). *Tascam 208i: Dream USB Interface for Podcasters & Streamers | GEAR REVIEW.* YouTube. https://www.youtube.com/watch?v=IF9kqn5hfGI&t=512s

Balanced headphones - What is that? Different from balanced cables? (2021, December 6). Adam Steel. https://youtu.be/7EQ3WbpibDs

Bass frequency surgery. (2017, April 19). Bob Katz. https://youtu.be/_ym-CAzJa_M

The best (And only) insulation material you need for DIY absorbers. (n.d.). Acoustics Insider. https://www.acousticsinsider.com/blog/best-insulation-material-diy-acoustic-absorbers

The best DIY interconnect. (2019, January 18). YouTube. https://youtu.be/Y_jkLCyJFPo

Best Live Sound Mic Placement Techniques. (2022, August 28). Sweetwater. https://www.sweetwater.com/insync/best-live-sound-mic-placement-techniques/

Better to use a power conditioner or straight into the wall? (2020, March 18). YouTube. https://youtu.be/RYXIWp3p9qA

Bilou. (2012, November 25). *MLV Bass Traps integrated in Ikea's Billy shelfs bottom.* Gearspace.com. https://gearspace.com/board/showpost.php?p=8478913&postcount=455

Booth Junkie. (2017, April 29). *Which pop filter should you use?* YouTube. https://youtu.be/amWbTkjfhDk

Booth Junkie. (2019, November 27). *416 Killer? Synco Mic D-2 Shotgun Mic for Voiceover.* YouTube. https://www.youtube.com/watch?v=NmJeMzl0JZI

Booth Junkie. (n.d.). *$1000 Mic Shootout - Sennheiser MKH416 vs Neumann TLM 103.* YouTube. https://www.youtube.com/watch?v=sdE_VekATvE

Bradly Hamilton. (n.d.). *What is Timbre? (2020)*. YouTube. http://www.youtube.com/watch?v=heiFFarVU6o

Branchus Creations. (n.d.). *Beginner's Guide to Soldering Electronics Part 1*. YouTube. https://www.youtube.com/watch?v=M2Jf8cebwCs

Brandt, J. N. (2010, September 8). *Width slats spacing sequence*. Gearspace.com. https://gearspace.com/board/showpost.php?p=5765376&postcount=25

A British Audiophile. (2020, May 16). *My Top 5 Inexpensive Hifi Tweaks*. YouTube. https://www.youtube.com/watch?v=Dtb88_hbCFQ&t=774s

The Broadcast Bridge. (2019, July 9). *The art of microphone placement*. https://www.thebroadcastbridge.com/content/entry/13629/the-art-of-microphone-placement

Brodie Brazil. (2021, September 19). *Five TRUTHS of the Sennheiser MKH 416*. YouTube. https://www.youtube.com/watch?v=swdJmrO7WiU

Budget but amazing musical paradise TUBE Dac? Musical paradise MP-D1 MK3 review. (2019, November 26). Thomas & Stereo. https://youtu.be/Da0gyq23lhs

Building the X-LS encore: Final assembly. (2020, August 17). YouTube. https://youtu.be/J_E4_CyHA5A

Cables make a difference - and we measured it! Alpha audio huge cable test. (2024, April 3). YouTube. https://youtu.be/8wVnURAckLI

Cadwell, B. (2023, March 9). *5 best invisible hearing aids in 2023 - Smallest and smartest*. Soundly. https://www.soundly.com/blog/best-invisible-hearing-aids

Capacitor safety - How to discharge capacitors safely. (2020, July 20).

YouTube. https://youtu.be/eQsfuS0VjJE?t=320

Chris Selim - Mixdown Online. (2019, February 20). *How to work with the dspMixFx in CUBASE and other DAWs.* YouTube. https://www.youtube.com/watch?v=eqW5hwFSqkA

Colt Capperrune. (2022, April 30). *Audio for Video - The Complete Guide.* YouTube. https://www.youtube.com/watch?v=hvDKeG63Qik

Colt Capperrune. (2022, May 4). *VOCAL EQ SECRETS of the PRO's.* YouTube. https://www.youtube.com/watch?v=Wq1di2luMcs

Construct your own audiophile power cable from £15 (no soldering). (2020, May 23). YouTube. https://youtu.be/gfVYfC6tmcc

Creative Sound Lab. (2018, June 6). *Pairing Two Very Different Mics on Guitar Cabinet.* YouTube. https://www.youtube.com/watch?v=VWc0HsQhpDU
Curtis Judd. (2012, January 11). *Review: Golden Age Project R1 Mark III Active Ribbon Mic.* YouTube. https://www.youtube.com/watch?v=haZzCqgQZ50

The dark side of the moon: Analog & digital comparison (CD, SACD, vinyl, tape). (2018, April 7). ANA[DIA]LOG. https://youtu.be/almdxI76DOw

Darko Audio. (2019, January 23). *A brief overview of the Dutch & Dutch 8c active loudspeaker.* YouTube. https://www.youtube.com/watch?v=fygm3iV_odA

DAW comparison chart | theDAWstudio.com. (n.d.). https://thedawstudio.com/resources/daw-comparison-chart/

Dawson, S. (2021, February 19). *How to get perfect sound from your DAC: Windows edition.* Addicted To Audio NZ. https://addictedtoaudio.co.nz/blogs/how-to/how-to-get-perfect-sound-from-your-dac-windows-edition

Denafrips Ares II DAC review. (2020, December 12).
YouTube. https://youtu.be/bb-O8JF24PI

Denon HEOS - Review and tutorial. (2019, December 13). HiDEF
Lifestyle. https://youtu.be/8Pd5KXWr7Hw

Diffraction in action. (2021, February 10). Alexander Mathematics
& Physics Tutoring. https://youtu.be/6cNrh92kcRU

DIY: Amp safety 101. (2014, July 22).
YouTube. https://youtu.be/DkEc58-vWc4

Do audiophile cables matter? Here's PROOF! (2020, June 29).
YouTube. https://youtu.be/DC0s6KqQz3g

Do preamps enhance sound quality? (2017, September 30).
YouTube. https://youtu.be/Dh27E7YKN9s

Do USB filters make DACs sound better? (2021, February 15).
YouTube. https://youtu.be/RulAcLrnPkA

*Dolby Atmos - What hardware and software do you need | Production
expert.* (2022, September 22). Production
Expert. https://www.pro-tools-expert.com/production-expert-
1/2020/5/30/dolby-atmos-what-hardware-and-software-do-i-need

Dolby Institute. (2021, September 14). *Dolby Atmos Music
Creation 101: Mixing for Headphones.*
YouTube. https://www.youtube.com/watch?v=qs1tffnAjPE

Dolby Professional. (n.d.). *Dolby Atmos music Panner.* Dolby
Professional - Create and Deliver Transformative Experiences in
Dolby - Dolby
Professional. https://professional.dolby.com/product/dolby-
atmos-content-creation/dolby-atmos-music-panner

Dolby Professional. (n.d.). *Dolby reference player.* Dolby

Professional - Create and Deliver Transformative Experiences in Dolby - Dolby Professional. https://professional.dolby.com/product/media-processing-and-delivery/drp—dolby-reference-player

Doug Ferrara. (2009, July 4). *Hearing is believing - The ultimate small mixing & mastering room.* Ethan Winer. https://youtu.be/dB8H0HFMylo

Electro-voice RE20. (n.d.). Recording Hacks.com. https://recordinghacks.com/microphones/Electro-Voice/RE20

Elliott, R. (n.d.). *Digital Signal Processing.* Elliott Sound Products. https://sound-au.com/articles/dsp.htm

Ephe - Residential Energy Efficiency. (2022, April 1). *Cheapest way of double glazing your Windows.* https://youtu.be/hgb-wL_VsGM

FireWalk. (2021, December 4). *How to Fix Ground Loop Noise, Hiss, Buzz, & Hum (Simple & Cheap!).* YouTube. https://www.youtube.com/watch?v=einxGsiuwso

FocalOnline. (2012, May 11). *Do-it-Yourself Acoustical Treatment: How to Build a Diffuser.* YouTube. https://www.youtube.com/watch?v=qFeM2uWuMZI

Frank Martin. (2021, June 13). *Apple Music and Dolby Atmos on a Mac.* YouTube. https://www.youtube.com/watch?v=GUblp6EVtJ0

Gapster. (2021, August 4). *Speaker Cable termination Turn your Diy Cables into Pro cables.* YouTube. https://www.youtube.com/watch?v=OEAKIdo0urQ

Gearspace Forum. (n.d.). *Common Gas Flow Resistivity numbers.* Just a moment... https://gearspace.com/board/studio-building-acoustics/625978-common-gas-flow-resistivity-numbers.html

Geoff Martin. (2020, April 21). *1: Introduction to Room Acoustics.* YouTube. https://www.youtube.com/watch?v=6Q0joik6E74&t=748s

Geoff the Grey Geek. (2017, August 9). *Helping you connect your AV equipment.* https://geoffthegreygeek.com/

GIK Acoustics. (2021, March 25). *Speaker Placement: How far from the wall should I place my speakers?* YouTube. https://www.youtube.com/watch?v=T10_MLGOBfc

GIK Acoustics. (2022, December 6). *Diffusion concepts explained - How acoustic diffusers work and which one is right for you.* https://youtu.be/1MtJOGXVZ1w

GR-Research. (2020, June 30). *Do Audiophile Cables Matter? Here's PROOF!* YouTube. https://www.youtube.com/watch?v=DC0s6KqQz3g&feature=youtu.be

GR-Research. (2020, May 20). *Scared of Soldering? This is how you do it!* YouTube. https://www.youtube.com/watch?v=mVcOWx7hQiQ

Hearing is believing - The ultimate small mixing & mastering room. (2009, July 4). YouTube. https://youtu.be/dB8H0HFMylo

HiDEF LIFESTYLE. (2019, December 14). *Denon HEOS - Review and Tutorial.* YouTube. https://www.youtube.com/watch?v=8Pd5KXWr7Hw

HiFi Collective. (2014, November 29). *Speaker Cable no3 - Neotech Cable & Banana Plugs.* YouTube. https://www.youtube.com/watch?v=zP5HE2WGCcw

HiFi Collective. (2018, November 3). *How To: Make Your Own 4 Strand Harmonics Litz Speaker Cable Using Audio Note Banana Plugs.* YouTube. https://www.youtube.com/watch?v=8V-m66BKd9U

Home Theater Geeks. (2016, August 26). *Home Theater Geeks 318: Mixing Atmos.* YouTube. https://www.youtube.com/watch?v=v8Xto_1AjQs

How do tubes and capacitors influence the sound of the MP-701 Mk2? (2023, May 17). YouTube. https://youtu.be/XxqtiDKJA0E

How to build your own acoustic panels (DIY). (2021, May 1). AcousticsFREQ.com: Your Acoustics & Noise Control Resource. https://acousticsfreq.com/how-to-build-your-own-acoustic-panels/

How to fix ground loop noise, hiss, buzz, & hum (Simple & cheap!). (2021, December 3). YouTube. https://youtu.be/einxGsiuwso

How to get the best out of Apple Music high resolution Lossless. (2021, July 29). Addicted To Audio NZ. https://addictedtoaudio.co.nz/blogs/how-to/how-to-get-the-best-out-of-apple-music-high-resolution-lossless

How to make a four Strand round braid - spiral - 4 Strand plait. (2023, July 20). YouTube. https://youtu.be/G-7f9F3MF7Q

How to: Make your own 4 Strand harmonics Litz speaker cable using audio note banana plugs. (2018, November 2). YouTube. https://youtu.be/8V-m66BKd9U

The best (And only) insulation material you need for DIY absorbers. Acoustics Insider. https://www.acousticsinsider.com/blog/best-insulation-material-diy-acoustic-absorbers

I hate the Schüt Freya tube preamp because it's so good. (2019, April 12). YouTube. https://youtu.be/LwMXMtSprxI

The importance of internal speaker wires. (2021, January 4). YouTube. https://youtu.be/KYLDGe9K9OM

Inglis, S. (2014, October 1). *How effective are portable vocal*

booths? Sound On
Sound. https://www.soundonsound.com/reviews/how-effective-
are-portable-vocal-booths

Introduction to room acoustics. (2020, April 20). geoff
martin. https://youtu.be/6Q0joik6E74

Is recorded music (mostly) too bright? (2021, May 30).
YouTube. https://youtu.be/bipYWEWNc6w

Is the FM300A output stage the best amplifier at all ? (English). (2023,
February 1). YouTube. https://youtu.be/KL346EfvKDw

Izotope Inc. (2012, October 11). *Reduce Hiss with RX 2 | iZotope
Tips From A Pro.* YouTube. https://www.youtube.com/watch?
v=lW0daAXB7OA

Izotope Inc. (2021, October 15). *RX 9 De-hum: Instantly Remove
Hum and Background Noise from Audio.*
YouTube. https://www.youtube.com/watch?v=iDRQXJmC6Eg

*Izotope RX Audio Rescue: Voice Over Enhanced, Repaired, Saved,
Recovered.* (n.d.). Facebook
Groups. https://www.facebook.com/groups/AudioRescue

Izotope RX users. (n.d.). Facebook
Groups. https://www.facebook.com/groups/RXusers%20

Jlafrenz. (n.d.). *DIY "Superchunk" Corner Bass Traps Tutorial.* The
Emotiva
Lounge. https://emotivalounge.proboards.com/thread/54838/s
uperchunk-
corner-bass-traps-tutorial

John Brandt. (n.d.). *Accurate & Efficient Acoustics.* Jhbrandt –
accoustic designer. https://jhbrandt.net/

John Brandt. (n.d.). *Design guidelines for adding slats to acoustic panels.*
Gearspace Forum. https://gearspace.com/board/showpost.php?

p=5765376&postcount=25

Johnmunstudios. (2020, June 14). *UAD Console App Overview.*
YouTube. https://www.youtube.com/watch?v=Axcvj5IvMpY

Just a moment... (n.d.). Just a
moment... https://researchgate.net/figure/Frequency-ranges-of-
several-musical-instruments-30_fig3_228446442

The Largest Collection of Ambisonic Recordings in One Place. (n.d.).
AMBISONIC SOUND
LIBRARY. https://library.soundfield.com/

Leisuretec Distribution Ltd. (2021, August 10). *Microphone Polar
Patterns: The Basics.* YouTube. https://www.youtube.com/watch?
v=keBa2ocQInI

Lifatec glass Toslink cables. (n.d.). Lifatec
USA. https://www.lifatec.com/toslink2.html

Logic Pro X Life. (2021, January 6). *How to Get Your Music on
Spotify.* YouTube. https://www.youtube.com/watch?
v=yPvkyMcNkio

MacProVideoDotCom. (2022, November 21). *Ozone 10 201:
Mastering With Ozone - Intro.*
YouTube. https://www.youtube.com/watch?v=YrH1a1RcnOY

Making high quality speaker cables with Canare 4S11. (2021, January
1). YouTube. https://youtu.be/snLExbyIJto

Max Langridge. (2022, June 10). *What is aptX HD and which
devices support it?* Pocket-
lint. https://www.pocket-lint.com/headphones/news/qualcomm
/142601-what-is-aptx-hd-and-which-devices-support-it

Microphones and miking instruments. (2021, August 18). ontoitmedia
Live
Toolkit. https://www.livetoolkit.com.au/guide/microphones

MidFi Guy. (2021, December 23). *SingXer SA-1 vs Jotunheim 2 vs A90 vs THX AAA 789 [Solid State Showdown Pt.1].* YouTube. https://www.youtube.com/watch?v=429AMOXdYZY

Mike Thornton. (2021, August 9). *How does AvidPlay compare with other music distribution services.* Production Expert. https://www.pro-tools-expert.com/production-expert-1/2020/2/11/how-does-avid-play-compare-with-other-music-distribution-services-like-cd-baby-distrokid-and-more-we-investigate

Mike Thornton. (2022, September 22). *Dolby Atmos - What hardware and software do you need | Production expert.* Production Expert. https://www.pro-tools-expert.com/production-expert-1/2020/5/30/dolby-atmos-what-hardware-and-software-do-i-need

Mixbus Tv. (2022, January 7). *How to Use Split EQ by Eventide - Masterclass.* YouTube. https://www.youtube.com/watch?v=9zFmyKiMRf4

Modal ringing and resonance. (2007, August 19). Ethan Winer. https://youtu.be/aHkAFSZmMk4

Mods for musical paradise MP-301 - Tubes • Canuck audio mart hifi and audio forum. (n.d.). Canuck Audio Mart Canada's Largest Online Hifi, Audio & Home Theater Classifieds. https://www.canuckaudiomart.com/forum/viewtopic.php?f=23&t=45210&start=75#p722668

Mogami studio gold XLR cable | Review & Teardown. (2021, November 25). YouTube. https://youtu.be/jWJ10EWQnTI

Mono and Stereo Cables, the difference between them. (n.d.). Gollihur Music. https://gollihurmusic.com/the-difference-between-mono-and-stereo-cables/

MotuTV. (2021, April 7). *CueMix 5.* YouTube. https://www.youtube.com/watch?v=aSWbuzE8Unc

MQA explained: Everything you need to know about high-res audio. (2017, May 2). Ars Technica. https://arstechnica.com/gadgets/2017/05/mqa-explained-everything-you-need-to-know-about-high-res-audio/

Music Tech Explained. (2021, November 9). *Logic Pro - What's new in 10.7 (With in-depth Dolby Atmos Explanations).* YouTube. https://www.youtube.com/watch?v=-WWhJQNo2zU

Musical Paradise Tube Dac MP-D1 MK3. (2019, November 26). Thomas & Stereo. https://youtu.be/Da0gyq23lhs

My top 5 inexpensive hifi tweaks. (2020, May 15). YouTube. https://youtu.be/Dtb88_hbCFQ?t=824

Mylesvphillips. (2012, May 17). *Bunbury backwash 2012.* YouTube. https://youtu.be/_9BvrcutD2E

Mynewmicrophone.com. (2020, February 14). *Top 11 best tube condenser microphones on the market 2023.* My New Microphone. https://mynewmicrophone.com/top-best-tube-condenser-microphones-on-the-market/

One moment, please... (n.d.). One moment, please... https://www.alpha-audio.net/review/interlinks-dont-do-anything-or-do-they-32-rca-cables-analyzed/

One of them gets close to my $3k reference?!? My fave speaker cables under $100! (2022, March 6). YouTube. https://youtu.be/44n1AEQO6nY?t=315

Paul McGowan, PS Audio. (2018, March 9). *How to easily setup a subwoofer.* YouTube. https://youtu.be/xIEmZA_ruIg

Paul McGowan, PS Audio. (2021, January 5). *The importance of internal speaker wires.* YouTube. https://www.youtube.com/watch?v=KYLDGe9K9OM

Paul McGowan, PS Audio. (2022, September 8). *DSD64, DSD512, and Octave Records mastering*. https://www.youtube.com/watch?v=W98Bs5t9oec

Paul McGowan, PS Audio. (2022, September 17). *How to play DSD*. https://youtu.be/ZUK9owGlqlY

Paul McGowan, PS Audio. (n.d.). *How important is solder?* YouTube. https://www.youtube.com/watch?v=mBB-ZxCdC2E

Pearl Acoustics. (2023, January 10). *Part 3: Analog vs digital - can vinyl ever sound better than digital?* https://youtu.be/MlccCTy4PiQ

Podcastage. (2019, February 27). *Sennheiser MKH-416 Short Shotgun Mic Review / Test*. YouTube. https://www.youtube.com/watch?v=hI3YW7DjOk8

Podcastage. (2020, November 11). *Townsend Labs Sphere L22 Review / Test (Compared to Neumann U87 Ai, U67, SM7b, & More)*. YouTube. https://www.youtube.com/watch?v=kGD4H-XkARE

Produce Like A Pro. (2021, April 26). *The 5 rules of mixing*. https://youtu.be/NE1Ov-ObLv0

Produce Like A Pro. (2022, September 5). *My favourite Plugins for mixing in 2022 - Building a home studio Pt. 10*. https://youtu.be/EDezID2RH6g

Production Expert. (2019, September 9). *Demo - Penteo 16 Pro - An All-in-one Upmix And Downmix Plugin Supporting Up To 16 Channels*. YouTube. https://www.youtube.com/watch?v=6AO48_B5GIA

Production Expert. (2021, June 23). *Using Synchro Arts VocAlign Ultra For ADR in Post*. YouTube. https://www.youtube.com/watch?v=Sbszd20p8-M
QRD diffusers: Technical reference. (n.d.). Collo's DIY Subwoofer Tools. https://www.subwoofer-builder.com/qrd.htm

Rebel Tech. (2020, September 15). *Greatest Microphone ever made? (Sennheiser MKH 416 Review and Test)*. YouTube. https://www.youtube.com/watch?v=yBqu3qKP2rk

Recording Hacks. (n.d.). *Electro-voice RE20*. https://recordinghacks.com/microphones/Electro-Voice/RE20

Recording Hacks. (n.d.). *Shure SM7B*. https://recordinghacks.com/microphones/Shure/SM7B

Recordingrevolution. (2016, October 17). *Mixing with EQ - Carving EQ holes*. https://youtu.be/uoJqNzxYQD4

Red Baarnes Audio. (2017, August 30). *Studio One PRIME-FREE Punch & Roll for Audiobooks & Voice Overs*. https://youtu.be/_9BvrcutD2E. https://www.youtube.com/watch?v=pzhWImZDlV0

Reisman, R. (2023, April 12). *Best hearing aids for your iPhone*. SeniorLiving.org. https://www.seniorliving.org/hearing-aids/best/iphone/

Reisong A12 modifications. (n.d.). SkunkieDesigns. https://www.skunkiedesigns.com/a12-mods

Review and measurements of Schiit WYRD USB filter. (2018, December 17). Audio Science Review (ASR) Forum. https://www.audiosciencereview.com/forum/index.php?threads/review-and-measurements-of-schiit-wyrd-usb-filter.5717/

Review: Elekit TU-8500 full-function preamp, part I. (n.d.). Doctorjohn Cheaptubeaudio: Audio Reviews and More. https://cheaptubeaudio.blogspot.com/2015/03/review-elekit-tu-8500-full-function.html

RME Audio. (2017, October 25). *RME Audio TotalMix FX - DSP Effects Overview*.

YouTube. https://www.youtube.com/watch?v=yfhNh4WmPso

Rogerson, B. (2022, October 19). *Best DAWs 2023: The best digital audio workstations for PC and Mac.*
MusicRadar. https://www.musicradar.com/news/the-best-daws-the-best-music-production-software-for-pc-and-mac

Room Modes And Harmonics Example - Room Acoustics. (2021, July 31). John Heisz - Speakers and Audio
Projects. https://www.youtube.com/watch?v=vr3kmMRWwBw

Room Modes vs. Acoustic Iinterference. (n.d.).
RealTraps.com. https://realtraps.com/art_modes.htm

Santi. (2014, October 14). *Corner bass traps DIY.*
HiFiVision.com. https://www.hifivision.com/threads/corner-bass-traps-diy-caution-picture-load.54553/

Scared of soldering? This is how you do it! (2020, May 19).
YouTube. https://youtu.be/mVcOWx7hQiQ

Schiit Lyr+ tube hybrid headphone amp - Super versatile, super fun! (2023, January 10).
YouTube. https://youtu.be/mDVH7QW_RGY

Schum, D. (n.d.). *Making speech more distinct.*
AudiologyOnline. https://www.audiologyonline.com/articles/making-speech-more-distinct-12469

The secret to good imaging. (2017, November 4).
YouTube. https://youtu.be/OI7DA886-9o

Sennheiser MKH-416 short shotgun mic review / Test. (2019, February 26). YouTube. https://youtu.be/hI3YW7DjOk8

Shure SM7B. (n.d.). Recording
Hacks. https://recordinghacks.com/microphones/Shure/SM7B

Shutterstock Tutorials. (2019, June 14). *How to Record ADR*

Dialogue | Filmmaking Tips.
YouTube. https://www.youtube.com/watch?v=7W46DjoFlGE

Slade Templeton. (2022, March 20). *Top 5 tube saturation Plugins: An audio showdown of the best dedicated tube saturators — SonicScoop.* SonicScoop. https://sonicscoop.com/top-5-tube-saturation-plugins-showdown-best-dedicated-tube-saturators/

Snug fit cover for bass trap. (2023, May 11). Warwick Thorn. https://youtu.be/wXlhGeI6hRA
Soundstagenetwork. (2021, January 30). *System Audio Legend 5 Silverback*

Active Loudspeakers Review! (Take 2, Ep:25).
YouTube. https://www.youtube.com/watch?v=E-pNJTZv06g

Stehlik's Music. (n.d.). *1:51 / 4:42 • Harmonic series Timbre - Sound Quality or Tone Color.* YouTube. https://www.youtube.com/watch?v=WihdMYEmol0

Studio One Narrators, VO & Podcasters. (n.d.). *Facebook.* Facebook
Groups. https://m.facebook.com/groups/StudioOneNarrationVO/

Sundaram, S. A. (2012, December 1). *What's wrong with digital volume controls?* Feature
Articles. https://www.soundstageultra.com/index.php/features-menu/general-interest-interviews-menu/311-what-s-wrong-with-digital-volume-controls

Tharbamar. (2021, January 2). *Making High Quality Speaker Cables with Canare 4S11.* YouTube. https://www.youtube.com/watch?v=snLExbyIJto

Thetubestore, S. @. (2023, August 25). *Audiophile review - 6L6 type vacuum tube Shootout.* thetubestore
Blog. https://blog.thetubestore.com/audiophile-review-6l6-type-vacuum-tube-shootout/

ThingMan. (2018, November 11). *Folded-well-Diffusers by ReadScapes*. ReadScapes. https://dngmns.home.xs4all.nl/fwd_uk.html

ThingMan. (n.d.). *Folded-well-Diffusers by ReadScapes*. ReadScapes by ThingMan. https://dngmns.home.xs4all.nl/fwd_uk.html

Think Media. (2021, December 17). *???? how to use your Sony as a WEBCAM! (HDMI or USB setup)*. https://youtu.be/fj3qxLRH-Po

The Three Techs. (2021, July 5). *How to use Sony ZV-1 as a webcam and settings*. https://youtu.be/C1Nvp_46ZO0

Tim Perry. (2015, July 6). *DIY sound diffusers FAQ*. Arqen.com. https://arqen.com/sound-diffusers/faq/

Top 10 best tube condenser microphones on the market 2024. (2024, January 4). My New Microphone. https://mynewmicrophone.com/top-best-tube-condenser-microphones-on-the-market/

Very nice musical paradise TUBE small integrated amp - MP 301 review. Great for tube rolling. (2019, July 3). YouTube. https://youtu.be/T05Rs9DiWUE

Vocal EQ cheat sheet: How to mix & EQ vocals (2024). (2024, March 15). NYC's top music production school: learn with certified trainers. https://343labs.com/vocal-eq-cheat-sheet/

Vocal EQ secrets of the pro's. (2022, May 4). YouTube. https://youtu.be/Wq1di2luMcs

Walker, R. (1990). *The Design and Application of Modular, Acoustic Diffusing Elements* (15). BBC Research Department Report. https://downloads.bbc.co.uk/rd/pubs/reports/1990-15.pdf

Warm vs. bright pianos: What's the difference? (2022, August 23). YouTube. https://youtu.be/KMd7wD_YpR4

Waves Audio. (2017, December 14). *Mixing in surround: DOs and don'ts.* waves.com. https://www.waves.com/mixing-in-surround-do-and-dont

What Hi-Fi?, W. (2021, October 5). *Cambridge audio DacMagic 200M review.* whathifi. https://www.whathifi.com/reviews/cambridge-audio-dacmagic-200m

What is the first reflection point? (n.d.). Advanced Acoustics. https://advancedacoustics-uk.com/pages/what-is-the-first-reflection-point

Which streaming service sounds the best? (2018, February 16). White Sea Studio. https://youtu.be/FURPQI3VW58

White Sea Studio. (2021, November 10). *Eventide just REVOLUTIONISED the EQ?!?!* YouTube. https://www.youtube.com/watch?v=7viYyRBmH7U

White Sea Studio. (2021, June 6). *Will these make EVERY STUDIO sound GREAT? || Dutch&Dutch 8C REVIEW.* YouTube. https://www.youtube.com/watch?v=aK-gdXCJoVg

Willsenton R300 integrated amp review. (2022, October 8). Steve Huff Photo & HiFi. https://youtu.be/BOcgEOecjPo

Yamaha Corporation of America. (2015, August 21). *Introducing MusicCast.* YouTube. https://www.youtube.com/watch?v=PPLrHbVEbCA

Z review - Schiit Jotunheim. (2016, October 23). https://youtu.be/vDvGNMUwOkI

Z Reviews. (2022, August 23). *Where do we go from here? || The*

Topping A90 Discrete ... A90D.
YouTube. https://www.youtube.com/watch?v=5nsYR-8zjQY

AUTHOR BIO

I am a teacher, and writer, and over the last ten years became interested in recording audio and then how to improve my music and TV listening experience. For recording, I started off building a recording studio and when we moved to a new house I, instead, built decor friendly mobile acoustic treatment, that I could take with me if we ever moved again. I also built four DIY upward firing speakers, Along the way, I undertook a huge learning curve. So, many times I found myself reading into great detail on some aspect without knowing how it fitted into the overall scheme. Finally, 10 years of research later, it started to fit into place. If I could write it up, maybe other people could benefit from an overall scheme of thinking about audio recording and hi-fi listening.

Also by Warwick Thorn

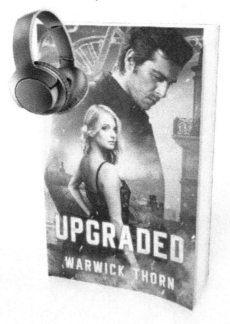